TSA Practice Papers

Volume Two

UniAdmissions

Copyright © 2018 *UniAdmissions*. All rights reserved.

ISBN 978-1-912557-44-8

No part of this publication may be reproduced or transmitted in any form or by any means, electronic or mechanical, including photocopying, recording, or by any information retrieval system without prior written permission of the publisher. This publication may not be used in conjunction with or to support any commercial undertaking without the prior written permission of the publisher.

Published by *RAR Medical Services Limited*
www.uniadmissions.co.uk
info@uniadmissions.co.uk
Tel: 0208 068 0438

This book is neither created nor endorsed by TSA. The authors and publisher are not affiliated with TSA. The information offered in this book is purely advisory and any advice given should be taken within this context. As such, the publishers and authors accept no liability whatsoever for the outcome of any applicant's TSA performance, the outcome of any university applications or for any other loss. Although every precaution has been taken in the preparation of this book, the publisher and author assume no responsibility for errors or omissions of any kind. Neither is any liability assumed for damages resulting from the use of information contained herein. This does not affect your statutory rights.

TSA Practice Papers

3 Full Papers & Solutions

Volume Two

Rohan Agarwal
Jonathan Madigan

UniAdmissions

About the Authors

Jon studied Economics and Management at St Hugh's College, Oxford, between 2013 and 2016. He sat the Thinkings Skills Assessment and **scored full marks** in section 1 of the paper, placing him in the **top 0.1% of candidates** who sat the assessment that year.

Jon has worked with UniAdmissions since 2014, working primarily as a personal tutor for a number of applicants across the entire application process. He has also provided Thinking Skills Assessment preparation courses within schools, and remains very familiar with all aspects of the paper.

Although Jon has moved into finance since graduation, he remains involved with tutoring as much as time allows. He regularly visits Oxford in his free time, and hopes to return for further study at some point in the future.

Rohan is the **Director of Operations** at *UniAdmissions* and is responsible for its technical and commercial arms. He graduated from Gonville and Caius College, Cambridge and is a fully qualified doctor. Over the last five years, he has tutored hundreds of successful Oxbridge and Medical applicants. He has also authored ten books on admissions tests and interviews.

Rohan has taught physiology to undergraduates and interviewed medical school applicants for Cambridge. He has published research on bone physiology and writes education articles for the Independent and Huffington Post. In his spare time, Rohan enjoys playing the piano and table tennis.

Introduction 6
   General Advice 8
   Revision Timetable 13
   Getting the most out of Mock Papers 14
   Before using this Book 15
   Section 1: An Overview 17
   Section 2: An Overview 18
   Section 2: Revision Guide 19
   Scoring Tables 22

Mock Papers 23
   Mock Paper D 23
   Mock Paper E 41
   Mock Paper F 53

Answer Key 67

Mock Paper Answers 68
   Mock Paper D: Section 1 68
   Mock Paper D: Section 2 76
   Mock Paper E: Section 1 80
   Mock Paper E: Section 2 86
   Mock Paper F: Section 1 90
   Mock Paper F: Section 2 98

Final Advice 102

Introduction

The Basics

The Thinking Skills Assessment is an aptitude test taken by students who are applying to certain courses at Cambridge and Oxford. Cambridge applicants sit the TSA Cambridge and Oxford applicants sit the TSA Oxford.

SECTION	SKILLS TESTED	QUESTIONS	TIMING
ONE	Problem-solving skills, including numerical and spatial reasoning. Critical thinking skills, including understanding argument and reasoning using everyday language.	50 MCQs	90 minutes
TWO	Ability to organise ideas in a clear and concise manner, and communicate them effectively in writing. Questions are usually but not necessarily medical.	One Essay from Four	30 minutes

NB: **TSA Oxford** consists of sections 1 + 2; **TSA Cambridge** and **TSA UCL** consist of only section 1.

Who has to sit the TSA?

Exam	Course
TSA Oxford	Students applying for the following subjects at Oxford **MUST** take the TSA: Economics and Management Experimental Psychology History and Economics Human Sciences Philosophy and Linguistics Philosophy, Politics and Economics (PPE) Psychology and Linguistics Psychology and Philosophy
TSA Cambridge	Students applying for Land Economy **MAY** have to take the TSA depending on the college they apply to.
TSA UCL	Students applying for European Social and Political Students **MUST** take the TSA.

NB: Applicants for Oxford **Chemistry or History & Economics** only have to complete section 1 of the TSA.

Preparing for the TSA

Before going any further, it's important that you understand the optimal way to prepare for the TSA. Rather than jumping straight into doing mock papers, it's essential that you start by understanding the components and the theory behind the TSA by using a TSA textbook. Once you've finished the non-timed practice questions, you can progress to past TSA papers. These are freely available online at www.uniadmissions.co.uk/tsa-past-papers and serve as excellent practice. You're strongly advised to use these in combination with the *TSA Past Worked Solutions* Book so that you can improve your weaknesses. Finally, once you've exhausted past papers, move onto the mock papers in this book.

Learn TSA Theory & Techniques
- Ultimate TSA Guide

→

Practice Questions
- TSA Past Paper Worked Solutions

→

TSA Mock Papers
- TSA Practice Papers

Already seen them all?

So, you've run out of past papers? Well hopefully that is where this book comes in. It contains six unique mock papers; each compiled by TSA Expert tutors at *UniAdmissions* and available nowhere else.

Having successfully gained a place at Oxford and scoring in the top 10% of the TSA, our tutors are intimately familiar with the TSA and its associated admission procedures. So, the novel questions presented to you here are of the correct style and difficulty to continue your revision and stretch you to meet the demands of the TSA.

General Advice

Start Early
It is much easier to prepare if you practice little and often. Start your preparation well in advance; ideally 10 weeks but at the latest within a month. This way you will have plenty of time to complete as many papers as you wish to feel comfortable and won't have to panic and cram just before the test, which is a much less effective and more stressful way to learn. In general, an early start will give you the opportunity to identify the complex issues and work at your own pace.

Prioritise
The MCQ section can be very time-pressured, and if you fail to answer the questions within the time limit you will be doing yourself a major disservice as every mark counts for this section. You need to be aware of how much time you're spending on each passage and allocate your time wisely. For example, since there are 50 questions in Section 1, and you are given 90 minutes in total, you will ideally take about 100 seconds per question (including reading time) so that you will not run out of time and panic towards the end.

Positive Marking
There are no penalties for incorrect answers; you will gain one for each right answer and will not get one for each wrong or unanswered one. This provides you with the luxury that you can always guess should you absolutely be not able to figure out the right answer for a question or run behind time. Since each question in Section 1 provides you with 5 possible answers, you have a 20% chance of guessing correctly. Therefore, if you aren't sure (and are running short of time), then make an educated guess and move on. Before 'guessing' you should try to eliminate a couple of answers to increase your chances of getting the question correct. For example, if a question has 5 options and you manage to eliminate 2 options- your chances of getting the question increase from 20% to 33%!

Avoid losing easy marks on other questions because of poor exam technique. Similarly, if you have failed to finish the exam, take the last ten seconds to guess the remaining questions to at least give yourself a chance of getting them right.

Practice
This is the best way of familiarising yourself with the style of questions and the timing for this section. Although the test does not demand any technical knowledge, you are unlikely to be familiar with the style of questions in all sections when you first encounter them. Therefore, you want to be comfortable at using this before you sit the test.

Practising questions will put you at ease and make you more comfortable with the exam. The more comfortable you are, the less you will panic on the test day and the more likely you are to score highly. Initially, work through the questions at your own pace, and spend time carefully reading the questions and looking at any additional data. When it becomes closer to the test, **make sure you practice the questions under exam conditions**.

Past Papers
Official past papers and answers are freely available online at **www.uniadmissions.co.uk/tsa-past-papers**. Practice makes perfect, and the more you practice the questions, especially for Section 1, the better you will get. Do not worry if you make plenty of mistakes at the start, the best way to learn is to understand why you have made certain mistakes and to not commit them again in the future!

You will undoubtedly get stuck when doing some past paper questions – they are designed to be tricky and the answer schemes don't offer any explanations. Thus, **you're highly advised to acquire a copy of *TSA Past Paper Worked Solutions*** – a free ebook is available online (see the back of this book for more details).

Repeat Questions
When checking through answers, pay particular attention to questions you have got wrong. If there is a worked answer, look through that carefully until you feel confident that you understand the reasoning, and then repeat the question without help to check that you can do it. If only the answer is given, have another look at the question and try to work out why that answer is correct. This is the best way to learn from your mistakes, and means you are less likely to make similar mistakes when it comes to the test. The same applies for questions which you were unsure of and made an educated guess which was correct, even if you got it right. When working through this book, **make sure you highlight any questions you are unsure of**, this means you know to spend more time looking over them once marked.

No Calculators and Dictionaries
The TSA requires a strong command of the English language, especially for Section 2 where you are asked to write an essay in 30 minutes. You are not allowed to use spell check or a dictionary, hence you should ensure that you written English is up to standard and you should ideally make close to no grammatical or spelling errors for your essay.

Section 1 contains several numerical reasoning questions, and you are not allowed to use a calculator, so make sure you are careful with your calculations.

Keywords
If you're stuck on a question, sometimes you can simply quickly scan the passage for any keywords that match the questions.

A word on Timing...

"If you had all day to do your exam, you would get 100%. But you don't."

Whilst this isn't completely true, it illustrates a very important point. Once you've practiced and know how to answer the questions, the clock is your biggest enemy. This seemingly obvious statement has one very important consequence. **The way to improve your score is to improve your speed.** There is no magic bullet. But there are a great number of techniques that, with practice, will give you significant time gains, allowing you to answer more questions and score more marks.

Timing is tight throughout – **mastering timing is the first key to success**. Some candidates choose to work as quickly as possible to save up time at the end to check back, but this is generally not the best way to do it. Often questions can have a lot of information in them – each time you start answering a question it takes time to get familiar with the instructions and information. By splitting the question into two sessions (the first run-through and the return-to-check) you double the amount of time you spend on familiarising yourself with the data, as you have to do it twice instead of only once. This costs valuable time. In addition, candidates who do check back may spend 2–3 minutes doing so and yet not make any actual changes. Whilst this can be reassuring, it is a false reassurance as it is unlikely to have a significant effect on your actual score. Therefore, it is usually best to pace yourself very steadily, aiming to spend the same amount of time on each question and finish the final question in a section just as time runs out. This reduces the time spent on re-familiarising with questions and maximises the time spent on the first attempt, gaining more marks.

It is essential that you don't get stuck with the hardest questions – no doubt there will be some. In the time spent answering only one of these you may miss out on answering three easier questions. If a question is taking too long, choose a sensible answer and move on. Never see this as giving up or in any way failing, rather it is the smart way to approach a test with a tight time limit. With practice and discipline, you can get very good at this and learn to maximise your efficiency. It is not about being a hero and aiming for full marks – this is almost impossible and very much unnecessary. It is about maximising your efficiency and gaining the maximum possible number of marks within the time you have.

Use the Options:

Some passages may try to trick you by providing a lot of unnecessary information. When presented with long passages that are seemingly hard to understand, it's essential you look at the answer options so you can focus your mind. This can allow you to reach the correct answer a lot more quickly. Consider the example below:

'Mountain climbing is viewed by some as an extreme sport, while for others it is simply an exhilarating pastime that offers the ultimate challenge of strength, endurance, and sacrifice. It can be highly dangerous, even fatal, especially when the climber is out of his or her depth, or simply gets overwhelmed by weather, terrain, ice, or other dangers of the mountain. Inexperience, poor planning, and inadequate equipment can all contribute to injury or death, so knowing what to do right matters.

Despite all the negatives, when done right, mountain climbing is an exciting, exhilarating, and rewarding experience. This article is an overview beginner's guide and outlines the initial basics to learn. Each step is deserving of an article in its own right, and entire tomes have been written on climbing mountains, so you're advised to spend a good deal of your beginner's learning immersed in reading widely. This basic overview will give you an idea of what is involved in a climb.'

Looking at the options first makes it obvious that certain information is redundant and allows you to quickly zoom in on certain keywords you should pick up on in order to answer the questions.

In other cases, **you may actually be able to solve the question without having to read the passage over and over again**. For example:

Which statement best summarises this paragraph?
A. *Mountain climbing is an extreme sport fraught with dangers.*
B. *Without extensive experience embarking on a mountain climb is fatal.*
C. *A comprehensive literature search is the key to enjoying mountain climbing.*
D. *Mountain climbing is difficult and is a skill that matures with age if pursued.*
E. *The terrain is the biggest unknown when climbing a mountain and therefore presents the biggest danger.*

If you read the passage first before looking at the question, you might have forgotten what the passage mentioned, and you will have to spend extra time going back to the passage to re-read it again.

You can **save a lot of time by looking at the questions first before reading the passage**. After looking at the question, you will know at the back of your head to look out for and this will save a considerable amount of time.

Manage your Time:

It is highly likely that you will be juggling your revision alongside your normal school studies. Whilst it is tempting to put your A-levels on the back burner falling behind in your school subjects is not a good idea, don't forget that to meet the conditions of your offer should you get one you will need at least one A*. So, time management is key!

Make sure you set aside a dedicated 90 minutes (and much more closer to the exam) to commit to your revision each day. The key here is not to sacrifice too many of your extracurricular activities, everybody needs some down time, but instead to be efficient. Take a look at our list of top tips for increasing revision efficiency below:

1. Create a comfortable work station
2. Declutter and stay tidy
3. Treat yourself to some nice stationery
4. See if music works for you → if not, find somewhere peaceful and quiet to work
5. Turn off your mobile or at least put it into silent mode
6. Silence social media alerts
7. Keep the TV off and out of sight
8. Stay organised with to do lists and revision timetables – more importantly, stick to them!
9. Keep to your set study times and don't bite off more than you can chew
10. Study while you're commuting
11. Adopt a positive mental attitude
12. Get into a routine
13. Consider forming a study group to focus on the harder exam concepts
14. Plan rest and reward days into your timetable – these are excellent incentive for you to stay on track with your study plans!

Keep Fit & Eat Well:

'A car won't work if you fill it with the wrong fuel' - your body is exactly the same. You cannot hope to perform unless you remain fit and well. The best way to do this is not underestimate the importance of healthy eating. Beige, starchy foods will make you sluggish; instead start the day with a hearty breakfast like porridge. Aim for the recommended 'five a day' intake of fruit/veg and stock up on the oily fish or blueberries – the so called "super foods".

When hitting the books, it's essential to keep your brain hydrated. If you get dehydrated you'll find yourself lethargic and possibly developing a headache, neither of which will do any favours for your revision. Invest in a good water bottle that you know the total volume of and keep sipping throughout the day. Don't forget that the amount of water you should be aiming to drink varies depending on your mass, so calculate your own personal recommended intake as follows: 30 ml per kg per day.

It is well known that exercise boosts your wellbeing and instils a sense of discipline. All of which will reflect well in your revision. It's well worth devoting half an hour a day to some exercise, get your heart rate up, break a sweat, and get those endorphins flowing.

Sleep

It's no secret that when revising you need to keep well rested. Don't be tempted to stay up late revising as sleep actually plays an important part in consolidating long term memory. Instead aim for a minimum of 7 hours good sleep each night, in a dark room without any glow from electronic appliances. Install flux (https://justgetflux.com) on your laptop to prevent your computer from disrupting your circadian rhythm. Aim to go to bed the same time each night and no hitting snooze on the alarm clock in the morning!

Revision Timetable

Still struggling to get organised? Then try filling in the example revision timetable below, remember to factor in enough time for short breaks, and stick to it! Remember to schedule in several breaks throughout the day and actually use them to do something you enjoy e.g. TV, reading, YouTube etc.

	8AM	10AM	12PM	2PM	4PM	8PM
TUESD						
WEDNES						
THURS						
FRIDAY						
SATUR						
SUNDAY						
EXAM DAY	School			Critical thinking		Essay

you have a much more accurate idea of the time you're spending on a question. In general, if you've spent >150 seconds on a section 1 question – move on regardless of how close you think you are to solving it.

Getting the most out of Mock Papers

Mock exams can prove invaluable if tackled correctly. Not only do they encourage you to start revision earlier, they also allow you to **practice and perfect your revision technique**. They are often the best way of improving your knowledge base or reinforcing what you have learnt. Probably the best reason for attempting mock papers is to familiarise yourself with the exam conditions of the TSA as they are particularly tough.

Start Revision Earlier

Thirty five percent of students agree that they procrastinate to a degree that is detrimental to their exam performance. This is partly explained by the fact that they often seem a long way in the future. In the scientific literature this is well recognised, Dr. Piers Steel, an expert on the field of motivation states that *'the further away an event is, the less impact it has on your decisions'*.

Mock exams are therefore a way of giving you a target to work towards and motivate you in the run up to the real thing – every time you do one treat it as the real deal! If you do well then it's a reassuring sign; if you do poorly then it will motivate you to work harder (and earlier!).

Practice and perfect revision techniques

In case you haven't realised already, revision is a skill all to itself, and can take some time to learn. For example, the most common revision techniques including **highlighting and/or re-reading are quite ineffective** ways of committing things to memory. Unless you are thinking critically about something you are much less likely to remember it or indeed understand it.

Mock exams, therefore allow you to test your revision strategies as you go along. Try spacing out your revision sessions so you have time to forget what you have learnt in-between. This may sound counterintuitive but the second time you remember it for longer. Try teaching another student what you have learnt; this forces you to structure the information in a logical way that may aid memory. Always try to question what you have learnt and appraise its validity. Not only does this aid memory but it is also a useful skill for the TSA, Oxbridge interviews, and beyond.

Improve your knowledge

The act of applying what you have learnt reinforces that piece of knowledge. An essay question in Section 2 may ask you about a fairly simple topic, but if you have a deep understanding of it you are able to write a critical essay that stands out from the crowd. Essay questions in particular provide a lot of room for students who have done their research to stand out, hence you should always aim to improve your knowledge and apply it from time to time. As you go through the mocks or past papers take note of your performance and see if you consistently under-perform in specific areas, thus highlighting areas for future study.

Get familiar with exam conditions

Pressure can cause all sorts of trouble for even the most brilliant students. The TSA is a particularly time pressured exam with high stakes – your future (without exaggerating) does depend on your result to a great extent. The real key to the TSA is overcoming this pressure and remaining calm to allow you to think efficiently.

Mock exams are therefore an excellent opportunity to devise and perfect your own exam techniques to beat the pressure and meet the demands of the exam. **Don't treat mock exams like practice questions – it's imperative you do them under time conditions.**

Before using this Book

Do the ground work
- Understand the format of the TSA – have a look at the TSA website and familiarise yourself with it: www.admissionstesting.org/thinking-skills-assessment
- Read widely in order to prepare yourself for the essay.
- Improve your written English if you are not confident in this aspect by practicing writing and reading frequently.
- Try to broaden your reading by learning about different topics that you are unfamiliar with as the essay topics can vary greatly.
- Learn how to understand a writer's viewpoint by reading news articles and having a go at summarising what the writer is arguing about.
- Be consistent – slot in regular TSA practice sessions when you have pockets of free time.
- Engage in discussion sessions with your friends and teachers – this might give you more ideas about certain essay topics.

Ease in gently
With the ground work laid, there's still no point in adopting exam conditions straight away. Instead invest in a beginner's guide to the TSA, which will not only describe in detail the background and theory of the exam, but take you through section by section what is expected. *The Ultimate TSA Guide* is the most popular TSA textbook – you can get a free copy by flicking to the back of this book.

When you are ready to move on to past papers, take your time and puzzle your way through all the questions. Really try to understand solutions. A past paper question won't be repeated in your real exam, so don't rote learn methods or facts. Instead, focus on applying prior knowledge to formulate your own approach.

If you're really struggling and have to take a sneak peek at the answers, then practice thinking of alternative solutions, or arguments for essays. It is unlikely that your answer will be more elegant or succinct than the model answer, but it is still a good task for encouraging creativity with your thinking. Get used to thinking outside the box!

Accelerate and Intensify

Start adopting exam conditions after you've done two past papers. Don't forget that **it's the time pressure that makes the TSA hard** – if you had as long as you wanted to sit the exam you would probably get 100%. If you're struggling to find comprehensive answers to past papers then *TSA Past Papers Worked Solutions* contains detailed explained answers to every TSA past paper question and essay (flick to the back to get a free copy).

Doing all the past papers is a good target for your revision. Choose a paper and proceed with strict exam conditions. Take a short break and then mark your answers before reviewing your progress. For revision purposes, as you go along, keep track of those questions that you guess – these are equally as important to review as those you get wrong.

```
        Do Paper under
           Exam
         Conditions
        ↗           ↘
Review questions      Mark Paper using
that you guessed      Answer Scheme
 or found hard
        ↖           ↙
         Review
      Questions that
       you got wrong
```

Once you've exhausted all the past papers, move on to tackling the unique mock papers in this book. In general, you should aim to complete one to two mock papers every night in the ten days preceding your exam.

Section 1: An Overview

What will you be tested on?	No. of Questions	Duration
Problem-solving skills, numerical and spatial reasoning, critical thinking skills, understanding arguments and reasoning	50 MCQs	90 Minutes

This is the first section of the TSA, comprising a total of 50 MCQ questions. You have 90 minutes in total to complete the MCQ questions, including reading time. In order to keep within the time limit, you realistically have about 108 seconds per question as you have to factor in the reading time as well.

Not all the questions are of equal difficulty and so as you work through the past material it is certainly worth learning to recognise quickly which questions you should spend less time on in order to give yourself more time for the trickier questions.

Deducing arguments
Several MCQ questions will be aimed at testing your understand of the writer's argument. It is common to see questions asking you 'what is the writer's view?' or 'what is the writer trying to argue?'. This is arguably an important skill you will have to develop, and the TSA is designed to test this ability. You have limited time to read the passage and understand the writer's argument, and the only way to improve your reading comprehension skill is to read several well-written news articles on a daily basis and think about them in a critical manner.

Assumptions
It is important to be able to identify the assumptions that a writer makes in the passage, as several questions might question your understanding of what is assumed in the passage. For example, if a writer mentions that 'if all else remains the same, we can expect our economic growth to improve next year', you can identify an assumption being made here – the writer is clearly assuming that all external factors remain the same.

Fact vs. Opinion
It is important to **be able to decipher whether the writer is stating a fact or an opinion** – the distinction is usually rather subtle and you will have to decide whether the writer is giving his or her own personal opinion, or presenting something as a fact. Section 1 may contain questions that will test your ability to identify what is presented as a fact and what is presented as an opinion.

Fact	Opinion
'There are 7 billion people in this world…'	'I believe there are more than 7 billion people in this world…'
'She is an Australian…'	'She sounded like an Australian…'
'Trump is the current President…'	'Trump is a horrible President…'
'Vegetables contain a lot of fibre…'	'Vegetables are good for you…'

Numerical and spatial reasoning
There are several questions that will test how well you can cope with numbers, and you should ideally be comfortable with simple mental calculations and being able to think logically.

Section 2: An Overview

What will you be tested on?	No. of Questions	Duration
Your ability to write an essay under timed conditions, your writing technique and your argumentative abilities	1 out of 4	30 Minutes

Section 2 is usually what students are more comfortable with – after all, many GCSE and A Level subjects require you to write essays within timed conditions. This section does not require you to have any particular specialist knowledge – the questions can be very broad and cover a wide range of topics.

Here are some of the topics that might appear in Section 2:

- Science
- Politics
- Religion
- Technology
- Ethics
- Morality
- Philosophy
- Education
- History
- Geopolitics

As you can see, this list is very broad and definitely non-exhaustive, and you do not get many choices to choose from (you have to write one essay out of three choices). Many students make the mistake of focusing too narrowly on one or two topics that they are comfortable with – this is a dangerous gamble and if you end up with four questions you are unfamiliar with, this is likely to negatively impact your score. **You should ideally focus on three topics to prepare from the list above**, and you can pick and choose which topics from the list above are the ones you would be more interested in. Here are some suggestions:

Science
An essay that is related to science might relate to recent technological advancements and their implications, such as the rise of Bitcoin and the use of blockchain technology and artificial intelligence. This is interrelated to ethical and moral issues, hence you cannot merely just regurgitate what you know about artificial intelligence or blockchain technology. **The examiners do not expect you to be an expert in an area of science** – what they want to see is how you identify certain moral or ethical issues that might arise due to scientific advancements, and how do we resolve such conundrums as human beings.

Politics
Politics is undeniably always a hot topic and consequently a popular choice amongst students. The danger with writing a politics question is that some **students get carried away and make their essay too one-sided or emotive** – for example a student may chance upon an essay question related to Brexit and go on a long rant about why the referendum was a bad idea. You should always remember to answer the question and make sure your essay addresses the exact question asked – do not get carried away and end up writing something irrelevant just because you have strong feelings about a certain topic.

Religion
Religion is always a controversial issue and essays on religion provide good students with an excellent opportunity to stand out and display their maturity in thought. Questions can range from asking about your opinion with regards to banning the wearing of a headdress to whether children should be exposed to religious practices at a young age. Questions related to religion will **require you to be sensitive and measured** in your answers - it is easy to trip up on such questions if you're not careful.

Education
Education is perhaps always a relatable topic to students, and students can draw from their own experience with the education system in order to form their opinion and write good essays on such topics. Questions can range from whether university places should be reduced, to whether we should be focusing on learning the sciences as opposed to the arts.

Section 2: Revision Guide

Science

Resource	What to read/do
1. Newspaper Articles	• The Guardian, The Times, The Economist, The Financial Times, The Telegraph, The New York Times, The Independent
2. A Levels/IB	• Look at the content of your science A Levels/IB if you are doing science subjects and critically analyse what are the potential moral/ethical implications • Use your A Levels/IB resources in order to seek out further readings – e.g. links to a scientific journal or blog commentary • Remember that for your LNAT essay you should not focus on the technical issues too much – think more about the ethical and moral issues
3. Online videos	• There are plenty of free resources online that provide interesting commentary on science and the moral and ethical conundrums that scientists face on a daily basis • E.g. Documentaries and specialist science channels on YouTube • National Geographic, Animal Planet etc. might also be good if you have access to them
4. Debates	• Having a discussion with your friends about topics related to science might also help you formulate some ideas • Attending debate sessions where the topic is related to science might also provide you with excellent arguments and counter-arguments • Some universities might also host information sessions for sixth form students – some might be relevant to ethical and moral issues in science
5. Museums	• Certain museums such as the Natural Science Museum might provide some interesting information that you might not have known about
6. Non-fiction books	• There are plenty of non-fiction books (non-technical ones) that might discuss moral and ethical issues about science in an easily digestible way

Politics

Resource	What to read/do
1. Newspaper Articles	• The Guardian, The Times, The Economist, The Financial Times, The Telegraph, The New York Times, The Independent
2. Television	• Parliamentary sessions • Prime Minister Questions • Political news
3. Online videos	• Documentaries • YouTube Channels
4. Lectures	• University introductory lectures • Sixth form information sessions
5. Debates	• Debates held in school • Joining a politics club
6. Podcasts	• Political podcasts • Listen to both sides to get a more rounded view (e.g. listening to both left and right wing podcasts)

Religion

Syllabus Point	What to read/do
2. Non-fiction books	› Read up about books that explain the origins and beliefs of different ty religion › E.g. Books that talk about the origins of Christianity, Islam or Buddhism, the books etc.
3. Talking to religious leaders	religions more and being able to write an essay on religion with more matur nuance › Talking to people from different religious backgrounds may also be a good forming a more well-rounded opinion
5. Lectures	› Information sessions › Relevant introductory lectures

Education

	What to read/do
1. Newspaper Articles	• The Guardian, The Times, The Economist, The Financial Times, T Telegraph, The New York Times, The Independent
2. A Levels/IB	you feel like what you are studying is useful and relevant? E.g. Studying a versus science • Compare the education you are receiving with your friends in differe schools or different subjects
4. University applications	• Have a read of how different universities promote themselves – do th claim to provide students with academic enlightenment, or better j prospects, or a good social life? • Why do different universities focus on different things?
6. Talk to your teachers	• Your teachers have been in the education industry for years and may decades – talk to them and ask them for their opinion • Talk to different teachers and compare their opinions regarding how

style rather than doing extra reading as the former will have a greater impact on your mark.

How to use this Book

If you have done everything this book has described so far then you should be well equipped to meet the demands of the TSA, and therefore **the mock papers in the rest of this book should ONLY be completed under exam conditions**.

This means:
- Absolute silence – no TV or music
- Absolute focus – no distractions such as eating your dinner
- Strict time constraints – no pausing half way through
- No checking the answers as you go
- Give yourself a maximum of three minutes between sections – keep the pressure up
- Complete the entire paper before marking
- Mark harshly

In practice this means setting aside 90 minutes for Section 1 and 30 minutes for Section 2 in an evening to find a quiet spot without interruptions and tackle the paper. Completing one mock paper every evening in the week running up to the exam would be an ideal target.

- Tackle the paper as you would in the exam.
- Return to mark your answers, but mark harshly if there's any ambiguity.
- Highlight any areas of concern.
- If warranted read up on the areas you felt you underperformed to reinforce your knowledge.
- If you inadvertently learnt anything new by muddling through a question, go and tell somebody about it to reinforce what you've discovered.

Finally relax… the TSA is an exhausting exam, concentrating so hard continually for 1.5 hours will take its toll. So, being able to relax and switch off is essential to keep yourself sharp for exam day! Make sure you reward yourself after you finish marking your exam.

Learn TSA Theory & Techniques	Practice Questions	TSA Mock Papers
• Ultimate TSA Guide	• TSA Past Paper Worked Solutions	• TSA Practice Papers

Top Tip! You can get free copies of *The Ultimate TSA Guide* and *TSA Past Paper Worked Solutions* by flicking to the back of this book.

Scoring Tables

Use these to keep a record of your scores from past papers – you can then easily see which paper you should attempt next (always the one with the lowest score).

	2nd Attempt	3rd Attempt
2008		
2010		
2012		
2015		

		2nd Attempt
Volume 1		
	Mock Paper C	
Volume 2		
	Mock Paper F	

You will not be able to give yourself a score for Section 2– the best way to gauge your performance for Section 2 will be to compare your arguments and counter-arguments with the model answer, or let your friends or teachers read it and gather some feedback from them. Fortunately for the mock papers in this book, there are model answers for you to compare your essays against!

Mock Papers

Mock Paper D

Question 1:
"Irish Folk Band, the Willow, have recently signed a contract with a new manager, and are organising a new musical tour. They and their manager are discussing which country would be best to organise their tour in. The lead singer of the Willow would like to organise a tour in Germany, which has a rich history of folk music. However, the manager finds that ticket sales for folk music concerts in Germany have been steadily declining for several years, whilst France has recently seen a significant increase in ticket sales for folk music concerts. The manager says that this means the group's ticket sales would be higher if they organise a tour in France, than if they organise one in Germany."

Which of the following is an assumption that the manager has made?
A. The band should prioritise profits and organise a tour in the most profitable country possible.
B. The band should not embark upon a new tour and should instead focus on record sales.
C. The decrease of ticket sales in Germany and the increase in France means that the band will sell fewer tickets in Germany than in France.
D. There will not be other countries which are even more profitable than France to organise the tour in.
E. Folk music is popular in France.

Question 2:
"Tom and Rob are going on a trekking holiday in Africa, organised by Wild Africa Inc. The holiday will include hiking, a safari day, kayaking and mountain climbing. Tom suffers from Asthma, and the pair does some research to find out if this will affect any part of their holiday. Rob is disappointed to find the following clause in the terms and conditions of Wild Africa Inc:
5.2 Customers suffering from severe Asthma, Breathing difficulties, Anaphylaxis or heart conditions are not allowed to take part in mountain-related activities on holidays organised by Wild Africa Inc.
Rob concludes that this means that Tom will not be allowed to take part in the mountain climbing section of their holiday."

Which of the following is an assumption that Rob has made?
A. The clause rob has found applies to their holiday.
B. Tom was intending to take part in the mountain climbing section of the holiday.
C. If Tom is not able to take part in the mountain climbing, Rob will not be able to either.
D. Tom suffers from severe asthma.
E. Wild Africa Inc. is following the correct health and safety regulations.

Question 3:
"Grace and Rose have both been attending an afterschool gymnastics class, which finishes at 5pm. After the class has finished, Grace and Rose cool down and change out of their gym clothes before heading home. Both girls depart at 5:15pm. Grace and Rose both live a 1.5 mile walk away from the local gymnasium. Therefore, they will definitely arrive home at the same time."

Which of the following is NOT an assumption made in this argument?
A. Both girls will walk at the same speed.
B. Both girls departed at the same time.
C. The gymnastics class is being held at the local gymnasium.
D. Grace will not get lost on the way home.
E. Both girls are walking home.

Question 4:
"John is a train enthusiast, who has been studying the directions in which trains travel after departing from various London Stations. He finds that trains departing from King's cross station in London head North on the East Coast Mainline, and travel to Edinburgh. Trains departing from Waterloo Station head West on the Southwest Mainline and travel to Plymouth. Trains departing from Victoria Station head South and travel to Kent.

John surmises that presently, in order to travel on a train from London to Edinburgh, he must get on at King's Cross Station."

Which of the following is an assumption that John has made?
A. The East Coast mainline has the fastest trains.
B. It would not be quicker to take a train from Waterloo to Southampton Airport, then travel to Edinburgh on an aeroplane.
C. Rail lines will not be built that will allow trains to travel from Waterloo Station or Victoria Station to Edinburgh.
D. Trains departing from King's Cross do not have any other destinations other than Edinburgh.
E. There are no other train stations in London from which trains may travel to Edinburgh.

Question 5:
"Modern cars have a many safety features such as seatbelts, crumple zones and enhanced braking systems. Though done in the name of safety, this gives many drivers a false sense of security, and causes them to drive much faster. This increase in driving speeds results in a great number of crashes at high speed, causing thousands of deaths each year as no amount of safety features is able to prevent deaths in the face of the immense forces involved in a crash at high speed. Therefore, we should install a new feature in cars, a spike on the steering wheel, pointing directly at the driver's chest. This would save thousands of lives each year, as when driving at lower speeds, drivers have much more control, and crashes are much more readily avoided."

Which of the following is an assumption made in this argument?
A. A spike in the steering wheel pointing at the driver's chest would cause people to drive more slowly.
B. Improving safety features in cars would not be sufficient to reduce deaths from high-speed crashes.
C. Crashes are much more readily avoided when driving at lower speeds.
D. Safety features have caused a lot of high-speed crashes.
E. High-speed crashes are causing millions of deaths each year.

Question 6:
"Tanks and armoured vehicles were a hugely influential factor in all battles in World War 2. German tanks were highly superior to the tanks used by France, and this was an essential reason why Germany was able to defeat France in 1940. However, Germany was later defeated in World War 2 by the Soviet Union. Germany lost a number of key battles such as the Battle of Stalingrad and the Battle of Kursk. These victories were essential for the eventual victory of the Soviet Union over Germany. Therefore, the Soviet Union's tanks in the battles of Stalingrad and Kursk must have been superior to those of Germany."

Which of the following is an assumption made in this argument?
A. Tanks were hugely influential in the Battle of Stalingrad.
B. The Battles of Stalingrad and Kursk were essential for the Soviet Union's victory over Germany.
C. The reasons why the Soviet Union defeated Germany in battle were the same as the reasons why Germany defeated France in battle.
D. German tanks being superior to those used by France was an essential reason why Germany was able to defeat France.
E. If the Soviet Union's tanks were superior to Germany's tanks, the Soviet Union's armoured vehicles must also have been superior to Germany's armoured vehicles.

~ 24 ~

Question 7:
"In the Battle of Waterloo, in 1815, French Emperor Napoleon Bonaparte's army was defeated by a British army commanded by British General Arthur Wellesley, Duke of Wellington. Essential to The British army's victory was the arrival of a group of Prussian reinforcements led by Field Marshal Von Blucher, which joined up with The British army and allowed them to overwhelm Bonaparte's left flank. Bonaparte had been aware of the threat posed by Von Blucher's Prussians, and had detached a force of French soldiers several days earlier under the command of Field Marshal Grouchy, with orders to engage the Prussians led by Von Blucher, and prevent them joining up with The British Army. However, whilst dining at a local inn, Grouchy mistook the sounds of gunfire for thunder, and believed that the battle had been cancelled. He therefore disobeyed his orders and did not engage the Prussians commanded by Von Blucher. Therefore, if Field Marshal Grouchy had not made this mistake and had engaged the Prussian force as commanded, the British would not have won the Battle of waterloo."

Which is the best statement of a flaw in this argument?
A. It implies that Field Marshal Grouchy was an incompetent commander, when in fact he was a highly respected general of the day.
B. It assumes that had Grouchy engaged the Prussian force, he would have been able to successfully prevent them joining up with the British army.
C. It assumes that the British army would not have been victorious without the arrival of the Prussian reinforcements.
D. It ignores the other mistakes made by Napoleon which contributed to the British army being victorious in the Battle of Waterloo.
E. It implies that thunder and gunshot sounds are frequently mistaken by generals.

Question 8:
"A cruise ship is sailing from Southampton to Barcelona, making several stops along the way at Calais and Bordeux, in France, Bilbao in Spain, and Porto in Portugal. At each stop, the ship must wait in a queue to be assigned a dock at which it can pull in, refuel and resupply. The busier the port, the longer the ship will have to queue to be assigned a dock. The Captain of the ship is planning the journey, and knows he must work out which ports will have the longest queues.
The Captain made the same journey last year, and found out that Bilbao was the busiest port in Europe during the course of the journey. He also knows that Bordeux is the busiest port in France, and that Porto is the busiest port in Portugal. Whilst he is planning the journey, he discovers that Calais is busier than Porto. The Captain concludes that he must plan for Bilbao to have the longest queue in the journey, Bordeux to have the second longest queue, Calais to have the third longest queue, and Porto to have the fourth longest queue."

Which of the following best illustrates a flaw in the Captain's reasoning?
A. Porto is less busy than Calais, but may be busier than Bordeux.
B. The rankings may have changed and Bilbao may no longer be the busiest port in Europe.
C. Just because a port is busier does not necessarily mean it will have the longest queues.
D. The ship may not have time to make all the stops.
E. The captain has forgotten to consider how many passengers will embark and disembark at each stop.

Question 9:
"Every year in Britain, thousands of children need to be taken to hospital due to unexpected Asthma attacks. Many of these attacks could be treated by inhalers, and not require hospitalisation, if inhalers were quickly available. However, in spite of this problem, very few schools stock inhalers. If the government enacted legislation requiring schools to keep a stock of inhalers, then many of these hospitalisations could be avoided."

Which of the following best illustrates a flaw in the reasoning of this argument?
A. It ignores the logistical problems associated with stocking inhalers in schools.
B. It assumes that inhalers could prevent asthma attacks as well as treating them.
C. It ignores the fact that inhalers can sometimes cause other medical problems due to allergic reactions.
D. It implies that if inhalers are stocked in schools they will be quickly available to children, in time to treat unexpected asthma attacks.
E. It neglects to consider asthma attacks which cannot be treated by an inhaler.

Question 10:
"The aeroplane was a marvel of modern engineering when it was first developed in the early 20th Century, and was testament to human ingenuity. Throughout the 20th Century, the aeroplane allowed humans to travel more freely and widely than ever before, and allowed people to see and appreciate the stunning natural beauty that the world has to offer. However, aeroplanes also produce lots of pollution, such as Carbon Dioxide and Sulphur Oxide. High levels of Carbon Dioxide in the atmosphere are currently causing global warming, which is destroying or damaging many natural environments throughout the world.
Therefore it is clear that the aeroplane, which once offered such opportunity to appreciate the world's natural beauty, has been largely responsible for damage to various natural environments throughout the world. We must now seek to curb air traffic in order to save the world's remaining natural environments."

Which of the following is the best statement of a flaw in this argument?
A. It assumes that aeroplanes are a major reason for the high levels of Carbon Dioxide in the atmosphere which are currently causing global warming.
B. It assumes that aeroplanes offer greater opportunity to appreciate the world's natural environments.
C. It assumes that high levels of Carbon Dioxide are responsible for global warming.
D. It does not consider the effects of Sulphur Dioxide pollution released by aeroplanes.
E. It implies that we should take action to prevent damage to the world's natural environments.

Question 11:
"Professors from the Department of Pathology at Amazonia University are conducting research into possible new treatments for malaria, which is caused by a microbe known as Plasmodium. Research from Sierra Leone, a third world country with a high rate of malaria, has found that liver cells in malaria patients are reactive to the antibody Tarpulin. Plasmodium is known to infect liver cells, and thus liver cells would react to Tarpulin if Plasmodium itself was reactive to Tarpulin. Thus the professors at Amazonia University begin to research how Tarpulin can be used to target Plasmodium and treat malaria. However, this research will not be successful, because liver cells would also react to Tarpulin if the wrong solution is used whilst conducting the experiments. Since malaria is not prevalent in Amazonia, the professors must rely on the data from Sierra Leone. If the experiments in Sierra Leone used the wrong solutions, then the liver cells would react to Tarpulin even if Plasmodium does not react to Tarpulin."

Which of the following best illustrates a flaw in this argument?
A. From the fact that Plasmodium infects liver cells, it cannot be inferred that infected liver cells would react to Tarpulin if Plasmodium does.
B. From the fact that the research was carried out in Sierra Leone, it cannot be inferred that the wrong solutions were used.
C. From the fact that the wrong solutions are used, it cannot be inferred that the liver cells would react to Tarpulin.
D. From the fact that Plasmodium is reactive to Tarpulin it cannot be assumed that Tarpulin can be used to combat Plasmodium.
E. From the fact that Liver cells react to Tarpulin, it cannot be inferred that Plasmodium is reactive to Tarpulin.

Question 12:
"Ancient Egypt was one of the world's most powerful nations for several thousand years, and built wondrous structures including the Sphinxes and the Great Pyramids to serve as a permanent reminder of its stature. Many other powerful nations throughout the ages have also built magnificent structures, such as the Colosseum built by the Romans, the Hanging Gardens of Babylon built by the Persians and the Great Wall of China built by the Chinese. As well as building magnificent structures, Rome, Persia and China had one other thing in common, namely a very strong military. Thus, history clearly shows us that in ancient times, for a nation to be a powerful nation, it must have had a very strong military. In addition to building great structures such as the pyramids, Ancient Egypt must have also possessed a very strong military."

Which of the following best illustrates the main conclusion of this argument?
A. In order to be a powerful nation, a nation must build magnificent structures.
B. In ancient times a very strong military was required to be a powerful nation.
C. Ancient Egypt built magnificent structures; therefore it must have been a powerful nation.
D. Rome, Persia and China were all powerful nations.
E. Ancient Egypt was a powerful nation; therefore, it must have had a very strong military.

Question 13:
Global warming is widely presented in modern society as a cause for significant concern. One particular area often thought to be at risk is the ice caps of the North and South Poles, which are often presented to be at risk of melting due to increased temperature. Environmentalist groups often campaign for energy consumption to be reduced, thus reducing CO_2 emissions, the leading cause of global warming. However, recent research shows that the North and South Poles are actually becoming cooler, not warmer, thanks to mysterious and unexplained weather patterns. Clearly, high energy consumption is not contributing to damage to the polar ice caps.

Which of the following best states a conclusion that can be drawn from this argument of this argument?
A. There is no point in reducing energy consumption for environmental reasons.
B. Reducing energy consumption will not reduce CO_2 emissions.
C. We should trust the recent research stating that the North and South poles are becoming cooler.
D. Reducing energy consumption will not contribute to saving the polar ice caps.
E. We should not be concerned about damage to the Polar Ice caps.

Question 14:
"In 1957 the drug Thalidomide was released, and used to relieve nausea and morning sickness during pregnancy. The pharmaceutical company which released Thalidomide had carried out extensive testing of the drug, and had carried out more tests than was required for new drugs in the 1950s. No adverse affects were reported, and the drug was thought to be safe and effective. However, after it was released, Thalidomide was found to be responsible for severe deformities in thousands of babies whose mothers had taken the drug whilst pregnant with them. When further research was carried out, it was found that the molecules in Thalidomide could adopt 2 molecular structures, known as Isotopes. One of these Isotopes was perfectly safe, but the other caused significant biological problems in pregnant women and had been responsible for the deformities in the babies. The company producing Thalidomide had not been aware of this 2nd Isotope when developing the drug."

Which of the following is a conclusion that can be drawn from this passage?
A. The company producing Thalidomide had acted irresponsibly by not carrying out the required level of testing for the drug.
B. No Isotopes of Thalidomide are safe.
C. The drug testing requirements in 1950s were not sufficient to identify all possible Isotopes of a given drug.
D. Thalidomide was not effective at relieving nausea and morning sickness.
E. The dangerous Isotope of Thalidomide was not effective at relieving nausea and morning sickness.

Question 15:
"As winter draws in and the weather turns colder, many bird species struggle to find enough places to live in modern cities, and populations are at risk of declining. The RSPB is working to tackle the problem, and has

pledged £500,000 towards building purpose-built homes for birds throughout many of the UK's major cities, with the aim of keeping bird populations as high as possible. However, the RSPB is taking a very narrow view of conservationism. The amount of money they have pledged should instead be used to purchase land and build mini nature reserves within major cities. Estimates suggest this would help a similar number of birds as the planned purpose-built homes. It would also offer the added benefit of helping many other types of wildlife, not just birds."

Which of the following best illustrates the main conclusion of the above passage?
A. The RSPB should seek to help as much wildlife as possible, not just birds.
B. The purpose-built homes for birds in major cities are a waste of money.
C. The estimates that building nature reserves would help as many birds as the planned purpose-built homes are correct.
D. The RSPB should build nature reserves instead of purpose-built homes in order to help the maximum amount of wildlife possible.
E. Nature reserved would help more wildlife than purpose-built homes for birds.

Question 16:
"The release of CO_2 from consumption of fossil fuels is the main reason behind global warming, which is causing significant damage to many natural environments throughout the world. One significant source of CO_2 emissions is cars, which release CO_2 as they use up petrol. In order to tackle this problem, many car companies have begun to design cars with engines that do not use as much petrol. However, engines which use less petrol are not as powerful, and less powerful cars are not attractive to the public. If a car company produces cars which are not attractive to the public, they will not be profitable."

Which of the following best illustrates the main conclusion of this argument?
A. Car companies which produce cars that use less petrol will not be profitable.
B. The public prefer more powerful cars.
C. Car companies should prioritise profits over helping the environment.
D. Car companies should seek to produce engines that use less petrol but are still just as powerful.
E. The public are not interested in helping the environment.

Question 17:
"Penicillin is one of the major success stories of modern medicine. Since its discovery in 1928, it has grown to become a crucial foundation of medicine, saving countless lives and introducing the age of antibiotics. Alexander Fleming is today given most of the credit for introducing and developing antibiotics, but in fact Fleming played a relatively minor role. Fleming initially discovered Penicillin, but was unable to demonstrate its clinical effectiveness, or discern ways of reliably and consistently producing it. 2 other scientists called Howard Florey and Ernst Chain were actually responsible for developing Penicillin to the point where it could be reliably produced and used in medicine, to treat infections in patients. Clearly, the credit for the wonders worked by Penicillin should not go to Fleming, but to Florey and Chain."

Which of the following best illustrates the main conclusion of this argument?
A. Fleming was unable to develop penicillin to the point of being a viable medical treatment.
B. The credit for Penicillin's effects on medicine should go to Ernst Chain and Howard Florey, not to Alexander Fleming.
C. Without Chain and Florey, Penicillin would not have been developed into a viable treatment.
D. Alexander Fleming only played a small role in the process of Penicillin becoming a feature of modern medicine.
E. Alexander Fleming is not given enough credit for his role in the development of penicillin.

Question 18:

"Barns and Co. wholesalers are seeking ways to increase staff productivity, and have commissioned research into methods that they can use to do this. One of their researchers reports that greater teamwork and co-operation between staff is well proven to increase productivity in businesses. She argues that group days out for staff, doing activities such as paintballing or boating, are proven to instil greater teamwork in staff.

One of the board members raises an objection, claiming that the expense and lost working days stemming from organising a staff day out will cause an immediate drop in productivity. The researcher states that this is indeed a downside to organising a staff day out, but argues that in the long term the increase in productivity from organising such an event will more than account for this short term drop."

What of the following is the best illustration of the main conclusion of the researcher's argument?

A. Barns and Co. should not organise a staff day out due to the immediate negative effect on productivity this will have.
B. The board should ignore the short term decrease in productivity that will result from organising a staff day out.
C. Barns and Co. must organise a staff day out or they risk a decrease in productivity.
D. The board member who questioned her research should be sacked.
E. The board of Barns and Co. should accept a short term decrease in productivity from organising a staff day out, in order to achieve the long-term increase in productivity that will result.

Question 19:

"Worcestershire Aquatic Centre is seeking to recruit a new dolphin trainer. They interview several candidates, and find that there are 2 candidates which are clearly more suitable than the others. They give both of these candidates a 2nd interview, with further questions about their experience and qualifications.

They discern that Candidate 1 has a proven capability to perform well to crowds, which is likely to bring in more profit to the Aquatic Centre as more people will come and watch the dolphin shows. However, Candidate 2 has more experience at handling dolphins, and a proven ability to do this successfully. The manager of the aquatic centre tells the recruiting officer that it is more important that the new dolphin trainer is able to successfully handle the dolphins, than that they are able to increase profit."

Which of the following best illustrates a conclusion that can be drawn from the above passage?

A. The recruiting officer will should hire Candidate 1 in order to increase profits at the Worcestershire Aquatic Centre.
B. The recruiting officer should ignore the manager, and hire whichever candidate he feels is the best.
C. The recruiting officer should hire Candidate 2, to ensure that the new trainer will be able to successfully handle the dolphins.
D. The recruitment officer should hire both candidates.
E. The 2 candidates are no more suitable than the others which were interviewed, and neither should be hired.

Question 20:

"Thrills and Spills Inc. is a theme park and water park in England. The park has Rollercoasters, Water Slides, Fun Rides, and Amusements. Currently, the park charges a fixed entry fee, and after customers have paid this they are able to access all rides free of charge. Stephen thinks this is unfair, and argues that it should be replaced by a system in which customers only pay for the rides you go on. Thus, only users who use Rollercoasters should pay for access to Rollercoasters, and likewise for the other types of ride."

Which of the following proposals does NOT follow the same reasoning that Stephen has used?

A. Every motorist is charged a flat rate of road tax, and the money raised is used towards the upkeep of all roads. This should be abolished, and replaced by a new system of toll roads. Then, only motorists who use motorways would have to pay towards their upkeep, and only motorists who use country roads would have to pay towards the upkeep of these roads.

B. Under the current system, everybody in the UK pays a flat rate of National Insurance tax, which is used to fund the NHS. This is unfair, and should be replaced by a system in which everybody pays for only the services that they use. For example, only those who require X-Rays should have to pay for the upkeep of the X-Ray departments in the NHS.

C. Communism is a terrible system of government, in that everybody has to pay for all public services regardless of which ones they use. A much better system is to privatise all public services, and make users pay, so that only people who use each service should pay for it.

D. Currently, everybody who watches TV pays a Television license, regardless of which channels they watch. This is unfair, and should be replaced by making all channels pay-to-view, so customers only have to pay for the channels they watch.

E. Currently everybody who uses the London Underground pays a flat rate for a day's access to the tube, regardless of how many journeys they make. This should be replaced by a system in which customers pay for each journey, and thus pay an amount dependent on how much they use the Tube.

Question 21:

"There has recently been a new election in the UK, and the new government is pondering what policy to adopt on the railway system in the UK. The Chancellor argues that the best policy is to have an entirely privatised railway system, which will encourage different train companies to be competitive, and try and attract customers by providing the best service at the lowest price, thus driving down costs and increasing quality for customers. However, the Transport Minister argues that this is a short-sighted policy. She argues that privatised companies will only run services on the most profitable lines, where there are lots of passengers. Under this system, train companies may choose not to run many services to rural areas. This will lead to rural communities being cut off and will cause issues of segregation and a lack of opportunities for people in these communities. She argues that public funding should be put towards rail services in order to ensure that people in rural communities are adequately served by rail services."

Which of the following, if true, would most strengthen the Transport Minister's argument?

A. The Transport Minister has ultimate power over railway policy, and she can overrule the Chancellor if she sees fit.

B. Many train services to rural communities currently have low passenger numbers, and are unlikely to be profitable.

C. French rail services receive high level of public funding, and users of these services enjoy good quality and low prices.

D. American railway services are privatised with no public funding, and yet rural communities in America are well served by railway services.

E. The Prime Minister agrees with the Transport Minister's line of argument. He sympathises with rural communities and does not believe in a privatised rail system.

Question 22:
"AIDEL supermarket is seeking to attract more customers, and gain a greater share of the local market. The board know that if they lower their prices, they will attract more customers. The board decides to lower prices across the store. Therefore they can expect their customer numbers to increase."

Which of the following most closely follows the reasoning in this paragraph?
A. Nisha is flying to America from London. If she flies from Gatwick airport, the price of the flight will be less than if she flies from Heathrow airport. Therefore, if she wishes to fly at a cheaper price than it will cost to fly from Heathrow, she must fly from Gatwick.
B. If a train is to arrive into Middlesbrough Station, it must first go through Thornaby Rail Station. At 18:53, a train pulls into Middlesbrough Station. Therefore, we know it must have passed through Thornaby Rail Station before 18:53.
C. Richard decided a long time ago that if he ever travels to London, he will ensure to go and watch Phantom of the Opera whilst he is there. For his 20th Birthday, Richard's parents pay for him to travel to London and spend several days there. Therefore, Richard will go and see Phantom of the Opera.
D. Red Squirrels are currently on the decline in the UK, mostly due to competition from Grey Squirrels. If the population of Red Squirrels is to increase, the number of Grey Squirrels must first decrease. The population of Grey Squirrels has recently decreased, and therefore we can expect to see an increase in the population of Red Squirrels.
E. If Norwich Rugby team are to win the East-of-England League, they must defeat Cambridge. Norwich are crowned champions of the East-of-England League. Therefore Norwich must have defeated Cambridge.

Question 23:
"The United States of America has a healthcare system which works on a pay-per-use basis, where users pay a certain amount dependent on how much they use the health service. This is ethically outrageous, as many citizens are too poor to pay for healthcare, and cannot therefore receive the treatment they need. It is much better to have a system where everybody pays an amount depending on how wealthy they are, such that the rich would pay more than the poor. This should then be used to fund the healthcare service, and treatment then provided to users based on need, and not on whether they can afford to pay for it."

Which of the following best illustrates the principle used in this argument?
A. There is shocking differentiation in the wealth of different people living in the world. Some live lives of vast extravagance, whilst others do not have the means to buy essentials. Therefore, we should tax the wealthy, and use the money to fund support for the poor.
B. The Police should not charge people who require help and assistance. Instead, everybody should pay a fixed amount, and the money used to fund the Police. The Police should then provide help and assistance to people based on need, not wealth.
C. Healthcare should be a universal right, so people should not have to pay for healthcare which they use. Instead, the health service should be paid for by taxing those responsible for major health conditions. For example, fast food chains should be taxed, as their products promote obesity, a major health concern. The Health Service should then provide care based on need, not wealth.
D. The Police should not be a publically funded body. Instead, those who require help from the police should pay for it, depending on how much help they require. This would encourage people to be less dependent on the police
E. The Fire Service should not charge people who require help with house fires. Instead, everybody should pay an amount depending on their income/personal wealth, and the money used to fund the Fire Service. The Fire Service should then provide help to people based on need, not wealth.

Question 24:
"The North York Moors is one of several National Parks in England. The Management team has been awarded a grant from the National Lottery and is looking for a way to attract more visitors to the Moors. Sam suggests that they invest in enhancing the natural landscapes present in the Moors, thus creating more beauty, and making people more inspired to visit. However, Lucy disagrees, and feels that they should invest in more visitor centres and information points. Lucy's argument is that whilst this will be more costly in terms of staffing these centres, the increase in visitor numbers will bring in more income for the Moors, and will counteract this extra cost."

Which of the following, if true, would most weaken Lucy's argument?
A. Visitor centres in other national parks do not generally generate as much revenue as they cost to staff.
B. National Lottery grants have a history of being badly spent by National Parks such as the North York Moors.
C. There are large numbers of people who are interested in volunteering to help the North York Moors and would be happy to staff visitor centres.
D. Another National Park, the Yorkshire Dales, has recently opened up 5 new visitor centres and seen their profits increase significantly.
E. The North York Moors is currently struggling to attract visitors.

Question 25:
"Recent statistics have found that in many important areas of current debate, the public are consistently incorrect in what they believe is happening. For example, the public believe that 24% of Benefit claims are fraudulent, when in fact 0.7% of benefit claims are fraudulent. The public also believe that Violent Crime has increased by 10% in the last 5 years, when it has actually fallen by 25%. Finally, the public believe that 5 times more money is spent on Jobseeker's allowance than on Pensions, when in fact 15 times more money is spent on Pensions than Jobseeker's allowance. Therefore, efforts to engage with the public on these issues of debate should be abandoned as it is pointless to try and argue with people who are so clearly wrong about the facts."

Which of the following best illustrates the principle used in this argument?
A. We should make efforts to increase public education, in order to ensure that the public are well informed about important matters.
B. It is important to ensure that the values of democracy are upheld. Therefore, when engaging on matters of medical policy, we should talk to all parties, even those which are often found to be incorrect about medical matters.
C. Kerry is looking to open up a new restaurant and is looking for advice. She should only talk to professional chefs, as these have the greatest expertise in this area.
D. There are many people who continue to campaign against vaccines, despite the fact that vaccines are one of the most beneficial discoveries of modern medicine. Research has shown time and time again that anti-vaccine campaigners are consistently wrong about how vaccines actually work. Therefore, there is no point in debating with anti-vaccine campaigners about vaccines.
E. Nobody knows everything, and everyone is incorrect on some matters. Therefore, we should seek to engage with as many people as possible on all matters.

Question 26:
When I board my train at York at 15:30, the announcer tells me it is 120 minutes to London Kings Cross. Assuming the announcement is accurate to the nearest 10 minutes and that the train is on time, what is the earliest time I might arrive at my destination which is a 10 minute walk from Kings Cross?
A. 17:20   B. 17:25   C. 17:30   D. 17:35   E. 17:40

Question 27:
My sister does 2 loads of washing per week plus an extra one for everyone who is living in the house that week. When her son is away at university, she buys a new carton of washing powder every 6 weeks, but when her son is home she has to buy a new one every 5 weeks. How many people are living in the house when her son is home?
A. 2   B. 3   C. 4   D. 5   E. 6

Question 28:
A train shuttle service runs between the city centre and the airport between 5:30am and 11:30pm on weekdays and 6:00am and midnight on weekends. There are two trains used to operate the service and each journey from the airport to the city centre or vice versa takes 24 minutes. It takes 4 minutes for the train to unload and 4 minutes to reload at the airport; 2 minutes for the train to unload and 2 minutes to reload at the city centre. What is the maximum number of journeys that can be made by the shuttle service in one day?

A. 36   B. 45   C. 60   D. 72   E. 90

Question 29:
Northern Line trains arrive into Waterloo Station every 6 minutes, Jubilee Line trains every 2.5 minutes and Bakerloo Line trains every 4 minutes. If trains from all 3 lines arrived into the station 4 minutes ago, how long will it be before they do so again?

A. 20 minutes
B. 26 minutes
C. 30 minutes
D. 54 minutes
E. 56 minutes

Question 30:
Sam is deciding whether to make her wedding invitations herself or get them professionally made at a cost of £1 each. She decides to work out how much it will cost to make them herself. Each invitation uses 1 sheet of cream card, 4 sheets of red paper and 1 metre of gold ribbon. She will also use a gold sticker on each invitation and stamp them with a stamper, which she will buy. The stamper needs a pad of ink which will last for 70 invitations. The table of stationery costs is shown below:

Product	Price
Red paper (pack of 100)	£2
15m roll of gold ribbon	£3
Pack of 30 gold stickers	£1
Stamper	£8
Ink pad	£4
Cream card (pack of 20)	£2

She wants to send 90 invitations and wants to have enough supplies for 4 spares only. How much will she save by making the invitations herself?

A. £15   B. £19   C. £23   D. £27   E. £31

Question 31:
Half of the boys in Mrs Nelson's class have brown eyes and two thirds of the class have brown hair. If we know that at least as many boys in the class as girls have brown hair, that there are at least as many boys than girls in the class and that there are 36 children in the class altogether, at least how many boys have both brown hair and brown eyes?

A. 2   B. 3   C. 4   D. 5   E. 6

Question 32:
Mandy is making orange squash for her daughter's birthday party. She wants to have a 300ml glass of squash for each of the 8 children attending and a 400ml glass of squash each for her and for 2 parents who are helping out. She has 600ml of the concentrated squash. What ratio of water:concentrated squash should she use in the dilution to ensure she has the right amount to go around?

A. 7:1   B. 6:1   C. 5:1   D. 4:1   E. 3:1

Question 33:
A group of students is picking A Level options from the table below. In each "block", they choose 1 option. They cannot choose the same option in more than one block, and they have to take a subject in every block. How many possible combinations of options are there?

Block A	Block B	Block C	Block D
English Literature	History	Biology	Chemistry
German	Physics	English Language	Economics
Mathematics	Psychology	Geography	Philosophy
Psychology	Spanish	Mathematics	Sociology

A. 16   B. 48   C. 64   D. 224   E. 256

Question 34:
Hannah opens a savings account for her 14 year old son who will go to university in 4 years time. She works out that he will need £19000 to live on at current prices so deposits this amount in the savings account when she opens it. The cost of living rises with inflation by 3% a year, and Hannah's money earns 4% a year on balances under £20000 or 5% a year on balances over £20000. The table below shows how much an increase of various percentages equates to for various amounts. The interest is calculated and paid into the account at the end of each year.

How much more money will Hannah's son have than he needs to live on to the nearest £100 (if he assumes the cost of living does not rise further and no further interest is earned after the start of his course in 4 years time)?

	3%	4%	5%
£19,000.00	£19,570.00	£19,760.00	£19,950.00
£19,570.00	£20,157.10	£20,352.80	£20,548.50
£19,760.00	£20,352.80	£20,550.40	£20,748.00
£19,950.00	£20,548.50	£20,748.00	£20,947.50
£20,157.10	£20,761.81	£20,963.38	£21,164.96
£20,352.80	£20,963.38	£21,166.91	£21,370.44
£20,548.50	£21,164.96	£21,370.44	£21,575.93
£20,550.40	£21,166.91	£21,372.42	£21,577.92
£20,748.00	£21,370.44	£21,577.92	£21,785.40
£20,761.81	£21,384.67	£21,592.29	£21,799.90
£21,372.42	£22,013.59	£22,227.31	£22,441.04
£21,577.92	£22,225.26	£22,441.04	£22,656.82

A. 800   B. 1000   C. 1300   D. 1400   E. 1500

Question 35:
Amaia needs to arrive at her university interview by 11am She has to travel from her house, a 10 minute walk from Southtown station, to Northtown University. Using the timetables below, what is the latest time she should she leave her house to arrive on time?

Train Timetable

Southtown	0913	0943	1013
Westtown	0924	0954	1024
Northtown West	0950	1020	1050
Northtown Central	0958	1028	1058
Easttown	1009	1039	1109

Bus Timetable

Northtown West station	959	1009	1019	1029
Northtown Shopping Centre	1009	1019	1029	1039
Northtown Central station	1020	1030	1040	1050
Northtown Football Club	1023	1033	1043	1053
Northtown University	1033	1043	1053	1103

A. 0903 B. 0913 C. 0923 D. 0933 E. 0943

Question 36:
Summer and Shaniqua are playing a game of "noughts and crosses". Each player is assigned either "noughts" (O) or "crosses" (X) and they take it in turns to choose an empty box of the 3x3 grid to put their symbol in. The winner is the first person to get a line of 3 of their symbol in any direction in the grid (vertically, horizontally or diagonally). Summer starts the game. The current position is shown below.

X	X	
	O	

Assuming Shaniqua now plays her symbol in the square which will stop Summer being able to win the game straight away, Summer should play in either of which 2 boxes to ensure she is able to win the game on the next turn no matter what Shaniqua does?

1	2	3
		4
6		5

A. 1 and 3 B. 1 and 5 C. 1 and 6 D. 2 and 4 E. 3 and 5

Question 37:
How many different squares (of either 1, 2, 3 or 4 grid squares in side length) can be made using the grid below?

A. 16	B. 20	C. 26	D. 28	E. 30

Question 38:
Dates can be written in a 6 digit format for example 12-10-14 would be 12th October 2014. How many days before this date was the last time when all 6 digits were different?

A. 11	B. 12	C. 13	D. 14	E. 15

Question 39:
A teacher is trying to arrange the 5 students in her class into a seating plan. Ashley must sit on the front row on the left hand side nearest the board because she has poor eyesight. Bella and Caitlin must not be sat in the same row as each other because they talk and disrupt the class. Danielle needs to be sat next to an empty seat as she sometimes has help from a teaching assistant. Emily should be sat on the end of a row because she has poor mobility and it is hard for her to get into a middle seat. The teacher has 2 tables which each sit 3 people, which are arranged 1 behind the other. Who is sitting in the front right seat?

A. Empty seat for the teaching assistant
B. Bella
C. Caitlin
D. Danielle
E. Emily

Question 40:
A packaging company wishes to make cardboard boxes by taking a flat 1.2m by 1.2m square piece of cardboard, cutting square sections out of each corner as shown by the picture below and folding up the sections remaining on each side to make a box. The company experiments with different size boxes by cutting differently sized squares from the corners each time. It makes a box with 10cm by 10cm squares cut out of each corner, a box with 20cm by 20cm squares cut out of each corner and so on up to one with 50cm by 50cm squares cut out of each corner.

Which side length cut out would result in a box with the largest volume?
A. 10 cm	B. 20 cm	C. 30 cm	D. 40 cm	E. 50 cm

Question 41:
Rental yield for buy-to-let properties is calculated by dividing the potential rent per year paid for a house by the amount it cost to buy the house and get it in a rentable condition. Tina is considering 5 houses as possible buy-to-let investments. House A is in good condition and could be rented as it is for £700 a month, and costs £168,000 to buy. House B is also in good condition but is a student house so Tina would need to buy furniture for it. The house would cost £190,000 to buy and £10,000 to furnish, but could be rented for 40 weeks of the year to 4 students at a rent of £125 a week each. House C needs a lot of work doing. It costs £100,000 but would need £44,000 of renovations, and would rent for £600 a month. House D costs £200,000 and would need £40,000 of renovations, and would rent out for £2000 a month. House E costs £80,000 and would need £20,000 of renovations, and could be rented out for £200 a week.

Which house has the highest rental yield?
A. A B. B C. C D. D E. E

Question 42:
Wendy is sending 50 invitations to her housewarming party by first class post. Every envelope contains an invitation weighing 70g, and some who are going to family and friends who live further away also contain a sheet of directions, which weigh 25g. The table below gives the prices of sending letters of certain weights by first or second class post. If the total cost of sending the invitations is £33, how many of the invitations contain the extra information?

	First Class	Second Class
Less than 50g	£0.50	£0.30
Less than 75g	£0.60	£0.40
Less than 100g	£0.70	£0.50
Less than 125g	£0.80	£0.60
Less than 150g	£0.90	£0.70

A. 15 B. 20 C. 25 D. 30 E. 35

Question 43:
If the mean of 5 numbers is 8, the median is 6 and the mode is 4, what must the two largest numbers in the set of numbers add up to?
A. 13 B. 16 C. 22 D. 26 E. 28

Question 44:
"St John's Hospital in Northumbria is looking to recruit a new consultant cardiologist, and interviews a series of candidates. The interview panel determines that 3 candidates are clearly more qualified for the role than the others, and they invite these 3 candidates for a second interview. During this second interview, and upon further examination of their previous employment records, it becomes apparent that Candidate 3 is the most proficient at surgery of the 3 candidates, whilst Candidate 1 is the best at patient interaction and explaining the risks of procedures. Candidate 2, meanwhile, ranks between the other 2 in both these aspects.

The hospital director tells the interviewing team that the hospital already has a well-renowned team dedicated to patient interaction, but the surgical success record at the hospital is in need of improvement. The director issues instructions that therefore, it is more important that the new candidate is proficient at surgery, and patient interaction is less of a concern."

Which of the following is a conclusion that can be drawn from the directors' comments?
A. The interviewing team should hire Candidate 2, in order to achieve a balance of good patient relations with good surgical records.
B. The interviewing team should hire Candidate 1, in order to ensure good patient interactions, as these are a vital part of a doctor's work.
C. The interviewing team should ignore the hospital director and assess the candidates further to see who would be the best fit.
D. The interviewing team should hire Candidate 3, in order to ensure that the new candidate has excellent surgical skills, to boost the hospital's success in this area.

Question 45:
A child's fare on the train is more than half but less than three quarters of the adult fare. Both adult and child fares are multiples of 20p. The cost for an adult and 2 children on the train is £4.20. How much is the difference between the adult fare and the child fare?
A. 60p   B. 80p   C. £1   D. £1.20   E. £1.40

Question 46:
A children's football tournament involves a group stage, then a knockout stage. In the group stage, 5 teams play in a round robin format (i.e. each team plays each other once) and the 2 who win the most matches proceed through to a knockout stage where 2 teams play each other and the one that wins proceeds to the next round until there are 2 teams left, who play the final. If we start with 40 teams, how many matches are played altogether?
A. 95   B. 96   C. 97   D. 98   E. 99

Question 47:
I write my 4 digit pin number down in a coded format, by multiplying the first and second number together, dividing by the third number than subtracting the fourth number. If my code is 3, which of these could my pin number be?
A. 3461   B. 9864   C. 5423   D. 7848   E. 6849

Question 48:
Niall can choose whether to join the gym or pay per session. Gym membership costs £30 a month and to attend classes or gym sessions as a member is free. Pay as you go gym sessions cost £4 and attending classes costs £2 each. Niall works out that it will cost him £2 more to pay per session than it will to buy membership. Which of these is a possible combination of Niall's gym sessions and classes for the month?
A. 5 gym sessions, 4 classes
B. 4 gym sessions, 4 classes
C. 5 gym sessions, 6 classes
D. 4 gym sessions, 6 classes
E. 5 gym sessions, 8 classes

Question 49:
Health and safety law dictates that I can only lift boxes of books that weigh in total up to half my weight. I have to move 120 books weighing 2kg each in boxes that when empty weigh 0.5kg. If I weigh 60kg, how many trips will I have to make to move all the books?
A. 5 B. 6 C. 7 D. 8 E. 9

Question 50:
Jenny joined her company 5 and a half years ago on a salary of £24000. At the end of every year of service her pay is reviewed and she gets a pay rise of 3y percent where y is the number of years she has been at the company, rounded up to the nearest £1000.

What is her salary now?
A. 36000 B. 37000 C. 38000 D. 39000 E. 40000

END OF SECTION

YOU MUST ANSWER **ONLY ONE** OF THE FOLLOWING QUESTIONS

Question 1:
Describe a scientific model of a human brain. Justify any materials you might need, how you would connect the separate components and highlight any limitations your model might have.

Question 2:
"Impatience is the driving force behind development."
　　　　　　　　　To what extent is patience no longer a virtue in modern society?

Question 3:
Voltaire once said, *"The perfect is the enemy of the good."*
　　　　　　　Discuss this claim.

Question 4:
Is world peace achievable?

END OF SECTION

Mock Paper E

Question 1:
A square sheet of paper is 20cms long. How many times must it be folded in half before it covers an area of 12.5cm^2?

A. 3   B. 4   C. 5   D. 6   E. 7

Question 2:
Mountain climbing is viewed by some as an extreme sport, while for others it is simply an exhilarating pastime that offers the ultimate challenge of strength, endurance, and sacrifice. It can be highly dangerous, even fatal, especially when the climber is out of his or her depth, or simply gets overwhelmed by weather, terrain, ice, or other dangers of the mountain. Inexperience, poor planning, and inadequate equipment can all contribute to injury or death, so knowing what to do right matters.

Despite all the negatives, when done right, mountain climbing is an exciting, exhilarating, and rewarding experience. This article is an overview beginner's guide and outlines the initial basics to learn. Each step is deserving of an article in its own right, and entire tomes have been written on climbing mountains, so you're advised to spend a good deal of your beginner's learning immersed in reading widely. This basic overview will give you an idea of what is involved in a climb.

Which statement best summarises this paragraph?
A. Mountain climbing is an extreme sport fraught with dangers.
B. Without extensive experience embarking on a mountain climb is fatal.
C. A comprehensive literature search is the key to enjoying mountain climbing.
D. Mountain climbing is difficult and is a skill that matures with age if pursued.
E. The terrain is the biggest unknown when climbing a mountain and therefore presents the biggest danger.

Question 3:
50% of an isolated population contract a new strain of resistant Malaria. Only 20% are symptomatic of which 10% are female. What percentage of the total population do symptomatic males represent?

A) 1%   B) 9%   C) 10%   D) 80%

Question 4:
John is a UK citizen yet is looking to buy a holiday home in the South of France. He is purchasing his new home through an agency. Unlike a normal estate agent, they offer monthly discount sales of up to 30%. As a French company, the agency sells in Euros. John decides to hold off on his purchase until the sale in the interest of saving money. What is the major assumption made in doing this?

A. The house he likes will not be bought in the meantime.
B. The agency will not be declared bankrupt.
C. The value of the pound will fall more than 30%.
D. The value of the pound will fall less than 30%.
E. The value of the euro may increase by up to 35% in the coming weeks.

Question 5:
In childcare professions, by law, there must be an adult to child ratio of no more than 1:4. Child minders are hired on a salary of £8.50 an hour. What is the maximum number of children that can be continually supervised for a period of 24 hours on a budget of £1,000?

A. 1 B. 8 C. 12 D. 16 E. 468

Question 6:
A table of admission prices for the local cinema is shown below:

	Peak	Off-peak
Adult	£11	£9.50
Child	£7	£5.50
Concession	£7	£5.50
Student	£5	£5

How much would a group of 3 adults, 5 children, a concession and 4 students save by visiting at an off-peak time rather than a peak time?

A. £11.50 B. £13.50 C. £15.50 D. £17.50 E. £18.50

Question 7:
All musicians play instruments. All oboe players are musicians. Oboes and pianos are instruments. Karen is a musician. Which statement is true?

A) Karen plays two instruments.
B) All musicians are oboe players.
C) All instruments are pianos or oboes.
D) Karen is an oboe player.
E) None of the above.

Question 8:
Flow mediated dilatation is a method used to assess vascular function within the body. It essentially adopts the use of an ultrasound scan to measure the percentage increase in the width of an artery before and after occlusion with a blood pressure cuff. Ultrasound scans are taken by one sonographer, and the average lumen diameter is then measured by an analyst. What is a potential flaw in the methodology of this technique?

A) Results will not be comparable within an individual if different arteries start at different diameters.
B) Results will not be comparable between individuals if they have different baseline arterial diameters.
C) Ultrasound is an outdated technique with no use in modern medicine.
D) This methodology is subject to human error.
E) This methodology is not repeatable.

Question 9:
If it takes 20 minutes to board an aeroplane, 15 minutes to disembark and the flight lasts two and a half hours. In the event of a delay it is not uncommon to add 20 minutes to the flight time. Megan is catching the flight in question as she needs to attend a meeting at 5pm. The location of the meeting is 15 minutes from the airport without traffic; 25 minutes with. Which of the following statements is valid considering this information?

A. If Megan wants to be on time for her meeting, given all possibilities described, the latest she can begin boarding at the departure airport is 1.30pm.
B. If Megan starts boarding at 1.40pm she will certainly be late.
C. If Megan aims to start boarding at 1.10pm she will arrive in time whether the plane is delayed or not.
D. If Megan wishes to be on time she doesn't have to worry about the plane being delayed as she can make up the time during the transport time from the arrival airport to the meeting.

Question 10:
A cask of whiskey holds a total volume of 500L. Every two and a half minutes half of the total volume is collected and discarded. How many minutes will it take for the entire cask to be emptied?

A) 80   B) 160   C) 200   D) 240   E) ∞

Question 11:
B is right of A. C is left of B. D is in front of C. E is in front of B. Where is D in relation to E?

A) D is behind E.
B) E is behind D.
C) D is to the right of E.
D) D is to the left of E.
E) E is to the left of D.

Question 12:
Car A has a fuel tank capacity of 30 gallons and achieves 40mpg. Car B on the other hand has a fuel tank capacity of 50 gallons but only achieves 30mpg. Both cars drive until they run out of fuel. If car A starts with a full tank of petrol and travels 200 miles further than car B, how full was car B's fuel tank?

A) 1/5   B) 1/4   C) 1/3   D) 1/2   E) 2/3

The information below relates to questions 13 and 14:

The art of change ringing adopts the use of 6 bells, numbered 1 to 6 in order of weight (1 being the lightest). Initially the bells are rung in this order: 1, 2, 3, 4, 5, 6 however the aim is to ultimately ring all the possible combinations of a 6-number sequence. The rules for doing this are very simple: each bell can only move a maximum of one place in the sequence every time it rings.

Question 13:
What is the total possible number of permutations of 6 bells?
A) 160   B) 220   C) 660   D) 720   E) 1160

Question 14:
Based on the information provided which of the following could be a possible series of bell sequences?

A. 1 2 3 4 5
   2 1 4 3 5
   2 3 1 5 4

B. 1 2 3 4 5
   2 1 4 3 5
   2 4 1 5 3

C. 1 2 3 4 5
   4 2 1 3 5
   4 2 3 1 5

D. 1 2 3 4 5
   1 4 3 2 5
   1 2 3 4 5

E. 1 2 3 4 5
   4 1 3 2 5
   5 3 1 2 4

Question 15:
The keypad to a safe comprises the digits 1 - 9. The code itself can be of indeterminate length. The code is therefore set by choosing a reference number so that when a code is entered the average of all the numbers entered must equal the chosen reference number.

Which of the following is true?

A) If the reference number was set greater than 9, the safe would be locked forever.
B) This safe is extremely insecure as if random digits were pressed for long enough it would average out at the correct reference number.
C) More than one number is always required to achieve the reference number.
D) All of the above are true.
E) None of the above are true.

Question 16:
The use of antibiotics is one of the major paradoxes in modern medicine. Antibiotics themselves provide a selection pressure to drive the evolution of antibiotic resistant strains of bacteria. This is largely due to the rapid growth rate of bacterial colonies and asexual cell division. As such a widespread initiative is in place to limit the prescription of antibiotics.
Which of the following is a fair assumption?

A) Antibiotic resistance is impossible to avoid as it is driven by evolution.
B) If bacteria reproduced at a slower rate antibiotic resistance would not be such an issue.
C) Medicine always creates more problems than it solves.
D) In the past antibiotics were used frivolously.
E) All of the above could be possible.

The information below relates to questions 17 – 21:
The Spaghetti Bolognese recipe below serves 10 people and each portion contains 300 kcal.

- 1kg mince
- 220g pancetta, diced
- 30g crushed garlic
- 1kg tinned tomatoes
- 300g diced onions
- 300g sliced mushrooms
- 200g grated cheese

Question 17:
What quantity of cheese is required to prepare a meal for 350 people?

A) 0.7kg   B) 7kg   C) 70kg   D) 700kg   E) 7000kg

Question 18:
If 12 portions represent 120% of an individual's recommended calorific intake, what is that individuals recommended calorific intake?

A) 2,600kcal   B) 2,800kcal   C) 3,000kcal   D) 3,200kcal   E) 3,400kcal

Question 19:
The recommended ratio of pasta to Bolognese is 4:1. If cooking for 30 people how much pasta should be used?

A) 30.3kg   B) 36.6kg   C) 42.9kg   D) 49.2kg   E) 55.5kg

Question 20:
What is the ratio of onions to the rest of the ingredients if garlic and pancetta are ignored?

A) 1/2.05   B) 1/3.9   C) 1/6.7   D) 1/9.3   E) 1/10

Question 21:
It takes 4 minutes to prepare the ingredients per portion, and a further 8 minutes per portion to cook. Simon has ample preparation space but is limited to cooking 8 portions at a time. What is the shortest period of time it would take him to turn all the ingredients into a meal for 25 people, assuming he didn't start cooking until all the ingredients were prepared?

A) 3 hours   B) 3 hours 40   C) 4 hours   D) 4 hours 40   E) 5 hours

Question 22:
A company sells custom design t-shirts. A breakdown of their costs is shown below:

Number of Items	Cost per Item — Black and white	Cost per Item — Colour
0 – 99	£3.00	£5.00
100 - 499	£2.50	£4.50
500 - 999	£2.00	£4.00
1000+	£1.00	£3.00

Customers with a never before printed design must also pay a surcharge of £50 to cover the cost of building a jig. What is the total cost for an order of unique stag do t-shirts: 50 in colour, and 200 in black and white?

A) £650   B) £700   C) £750   D) £800   E) £850

Question 23:
The Scouts is a movement for young people first established by Lord Baden Powell. As the founder he was the first chief scout of the association. Since his initial appointment there have been a number of notable chief scouts including Peter Duncan and Bear Grylls. Some of the first camping trips conducted by Lord Powell's scout troop were on Brown Sea Island.

Now the Scout movement is a worldwide global phenomenon giving children from all backgrounds the opportunity not only to embark upon adventure but also to engage in the understanding and teaching of foreign culture. Traditionally religion formed the back bone of the scouting movement which was reflected in the scouts promise: "I promise to do my duty to god and to the queen".

Which of the following applies to the scout movement?

A) Scouts work for the Queen.
B) The scout network is aimed at adventurous individuals.
C) Chief scout is appointed by the Queen.
D) You have to be religious to be a scout.
E) None of the above.

Question 24:
Three rats are placed in a maze that is in the shape of an equilateral triangle. They pick a direction at random and walk along the side of a triangle. Sophie thinks they are less likely to collide than not. Is she correct?

A) Yes, because mice naturally keep away from each other.
B) No, they are more likely to collide than not.
C) No, they are equally likely to collide than not collide.
D) Yes, because the probability they collide is 0.25.
E) None of the above.

Question 25:
The use of human cadavers in the teaching of anatomy is hotly debated. Whilst many argue that it is an invaluable teaching resource, demonstrating far more than a text book can. Others describe how it is an outdated method which puts unfair stress on an already bereaved family. One of the biggest pros for using human tissue in anatomical teaching is the variation that it displays. Whilst textbooks demonstrate a standard model averaged over many 100s of specimens, many argue that it is the variation between cadavers that really reinforces anatomical knowledge.

The opposition argues that it is a cruel process that damages the grieving process of the effected family. For the use of the cadaver often occupies a period of up to 12 months. As such the relative in question is returned to the bereaved family for burial around the time it would be expected that they were recovering as described in the grieving model.

Does the article support or reject the use of cadavers in anatomical teaching?

A) Supports the use
B) Rejects the use
C) Impartial
D) Can't tell
E) None of the above

Question 26:
A ferry is carrying its full capacity. At the time of departure (7am) the travel time to the nearest hour is announced as 13 hours. What is the latest that the ferry could arrive at its destination?

A) 08.00   B) 20.00   C) 20.29   D) 20.30   E) 20.50

Question 27:
A game is played using a circle of 55 stepping stones. A die is rolled showing the numbers 1 - 6. The number on the die tells you how many steps you may take during your go. The only rule is that during your go you must take your steps in the routine two steps forward, 1 step back.

What is the minimum number or rolls required to win?

A) 28   B) 55   C) 110   D) 165   E) 200

Question 28:
On a race track there are 3 cars recording average lap times of 40 seconds, 60 seconds, and 70 seconds. They all started simultaneously 4 minutes ago. How much longer will the race need to continue for them to all cross the start line again at the same time?

A. 23.33 hours   B. 46.67 hours   C. 60.00 hours   D. 83.33 hours   E. 106.67 hours

Question 29:
A class of 60 2nd year medical students are conducting an experiment to measure the velocity of nerve conduction along their radial arteries. This work builds on a previous result obtained demonstrating the effects of how right-handed men have faster nerve conduction velocities than gender matched left handed individuals. 60% of the class are female of which 3% were unable to take part due to underlying heart conditions. 2 of the male members of the class were also unable to take part. On average the female cohort had faster nerve conduction velocities than men in their dominant arm.

Right handed women have the fastest nerve conduction velocities.

A) True   B) False   C) Can't tell

Question 30:
Mark is making a double tetrahedron dice by joining two square based pyramids together at their bases. Each square based pyramid is 5cm wide and 8cm tall. What area of card would have been required to produce the nets for the whole die?

A) 150cm^2 B) 180 cm^2 C) 210 cm^2 D) 240 cm^2 E) 270 cm^2

Question 31:
A serial dilution is performed by lining up 10 wells and filling each one with 9ml of distilled water. 1 ml of a concentrated solvent is then added to the first well and mixed. 1 ml of this new solution is drawn from the first well and added to the second and mixed. The process is repeated until all 10 wells have been used.
If the solvent starts off at concentration x, what will its final concentration be after 10 wells of serial dilution?

A. $x/10^9$ B. $x/10^{10}$ C. $x/10^{11}$ D. $x/10^{12}$ E. $x/10^{13}$

Question 32:
A student decides to measure the volume of all the blood in his body. He does this by injecting a known quantity of substrate into his arm, waiting a period of 20 minutes, then drawing a blood sample and measuring the concentration of the substrate in his blood.

What assumption has he made here?

A) The substrate is only soluble in blood.
B) The substrate is not bioavailable.
C) The substrate is not excreted.
D) The substrate is not degraded.
E) All of the above.

Question 33:
Jason is ordering a buffet for a party. The buffet company can provide a basic spread at £10 per head. However more luxurious items carry a surcharge. Jason is particularly interested in cup-cakes and shell fish. With these items included the buffet company provides a new quote of £10 per head. In addition to simply ordering the food Jason must also purchase cutlery and plates. Plates come in packs of 20 for £8 whilst cutlery is sold in bundles of 60 sets for £10.

With a budget of £2,300 (to the nearest 10 people) what is the maximum number of people Jason can provide food on a plate for?

A) 180   B) 190   C) 209   D) 210   E) 220

Question 34:
What were once methods of hunting have now become popular sports. Examples include archery, the javelin throw, the discuss throw and even throwing a boomerang. Why such dangerous hobbies have begun to thrive is now being investigated by social scientists. One such explanation is that it is because they are dangerous we find them appealing in the first place. Others argue that it is a throwback to our ancestral heritage, where as a hunter gatherer being a proficient hunter was something to show off and flaunt. Whilst this may be the case it is well observed that many find the chase of a hunt exciting if not controversial.

Sports like archery provide excitement analogous to that of the chase during a hunter gatherer hunt.

A) True
B) False
C) Can't tell

Question 35:
"If vaccinations are now compulsory because society has decided that they should be forced, then society should pay for them." Which of the following statements would weaken the argument?

A) Many people disagree that vaccinations should be compulsory.
B) The cost of vaccinations is too high to be funded locally.
C) Vaccinations are supported by many local communities and GPs.
D) Healthcare workers do not want vaccinations.
E) None of the above

Question 36:
Josh is painting the outside walls of his house. The paint he has chosen is sold only in 10L tins. Each tin costs £4.99. Assuming a litre of paint covers an area of 5m^2, and the total surface area of Josh's outside walls is 1050m^2; what is the total cost of the paint required if Josh wants to apply 3 coats?

A) £104.79 B) £209.58 C) £314.37 D) £419.16 E) £523.95

Question 37:
The stars of the night sky have remained unchanged for many hundreds of years, which allows sailors to navigate using the North Star still to this day. However, this only applies within the northern hemisphere as the populations of the southern hemisphere are subject to an alternative night sky.

An asterism can be used to locate the North Star, it comes by many names including the plough, the saucepan, and the big dipper. Whilst the North Star's position remains fixed in the sky (allowing it to point north reliably always) the rest of the stars traverse around the North Star in a singular motion. In a very long time, the North Star will one day move from its location due to the movement of the Earth.

Which of the following is **NOT** an assumption made in this argument?

A) The Earth is rotating on its axis.
B) Sailors still have need to navigate using the stars.
C) An analogous southern star is used to navigate in the Southern hemisphere.
D) The plough is not the only method of locating the North Star.
E) None of the above.

Question 38:
John wishes to deposit a cheque. The bank's opening times are 9am until 5pm Monday to Friday, 10am until 4pm on Saturdays, and the bank is closed on Sundays. It takes on average 42 bank hours for the money from a cheque to become available.

If John needs the money by 8pm Tuesday, what is the latest he can cash the cheque?

A) 5pm the Saturday before
B) 5pm the Friday before
C) 1pm the Thursday before
D) 1pm the Wednesday before
E) 9am the Tuesday before

Question 39:
How many different diamonds are there in the image shown to the right?

A) 25
B) 32
C) 48
D) 58
E) 63

MOCK PAPER E — SECTION ONE

Question 40:
In 4 years time I will be one third the age that my brother will be next year. In 20 years time he will be double my age. How old am I?

A) 4   B) 9   C) 15   D) 17   E) 23

Question 41:
Aneurysmal disease has been proven to induce systemic inflammatory effects, reaching far beyond the site of the aneurysm. The inflammatory mediator responsible for these processes remains unknown, however the effects of systemic inflammation have been well categorised and observed experimentally in pig models.

This inflammation induces an aberration of endothelial function within the inner most layer of blood vessel walls. The endothelium not only represents the lining of blood vessels but also acts as a transducer converting the haemodynamic forces of blood into a biological response. An example of this is the NO pathway, which uses the shear stress induced by increased blood flow to drive the formation of NO. NO diffuses from the endothelium into the smooth muscle surrounding blood vessels to promote vasodilatation and therefore acts to reduce blood flow.

Failure of this process induces high risk of vascular damage and therefore cardiovascular diseases such as thrombosis and atherosclerosis.

What is a valid implication from the text above?

A) Aneurysmal disease does not affect the NO pathway.
B) Aneurysms directly increase the likelihood of cardiovascular disease.
C) Aneurysms are the opposite of transducers.
D) Observations of this kind should be made in humans to see if the results can be replicated.
E) Aneurysms induce high blood flow.

Question 42:
A traffic surveyor is stood at a T-junction between a main road and a side street. He is only interested in traffic leaving the side street. He logs the class of vehicle, the colour and the direction of travel once on the main road. During an 8-hour period he observes a total of 346 vehicles including bikes. Of which 200 were travelling west whilst the rest travelled east. The over whelming majority of vehicles seen were cars at 90% with bikes, vans and articulated lorries together comprising the remaining 10%. Red was the most common colour observed whilst green was the least. Black and white vehicles were seen in equal quantities.

Which of the following is an accurate inference based on his survey?

A) Global sales are highest for those vehicles which are coloured red.
B) Cars are the most popular vehicle on all roads.
C) Green vehicle sales are down in the area that the surveyor was based.
D) The daily average rate of traffic out of a T junction in Britain is 346 vehicles over 8 hours.
E) To the east of the junction is a dead end.

Question 43:
William, Xavier, and Yolanda race in a 100m race. All of them run at a constant speed during the race. William beats Xavier by 20m. Xavier beats Yolanda by 20m. How many metres does William beat Yolanda?

A) 30m   B) 36m   C) 40m   D) 60m   E) 64m

Question 44:
A television is delivered in a box that has volume 60% larger than that of the television. The television is 150cm x 100cm x 10m. How much surplus volume is there?

A) 0.09m^2 B) 0.9 m^2 C) 9 m^2 D) 90 m^2 E) 900 m^2

Question 45:
Matthew and David are deciding where they would like to go camping Friday to Sunday. Upon completing their research, they discover the following:
➢ Whitmore Bay charges £5.50 a night and does not require a booking. The site provides showers, washing up facilities and easy access to a beach
➢ Port Eynon charges £5 a night and a booking is compulsory. However, the site does not provide showers but does have 240V sockets free of charge
➢ Jackson Bay charges £7 a night and is billed as a luxury site with compulsory booking, private showers, toilets, mobile phone charging facilities and kitchens.

David presents the following suggestion:
As Port Eynon is the farthest distance to travel the benefit of its cheap nightly rate is negated by the cost of petrol. Instead he recommends they visit Jackson Bay as it is the shortest distance to travel and will therefore be the cheapest.

Which of the following best illustrates a flaw in this argument?

A) Whitmore bay may be only a few miles further which means the total cost would be less than visiting Jackson Bay.
B) With kitchen facilities available they will be tempted to buy more food increasing the cost.
C) The campsite may be fully booked.
D) There may be a booking fee driving the cost up above that of the other campsites.
E) All of the above.

Question 46:

The manufacture of any new pharmaceutical is not permitted without scrupulous testing and analysis. This has led to the widespread, and controversial use of animal models in science. Whilst it is possible to test cytotoxicity on simple cell cultures, to truly predict the effect of a drug within a physiological system it must be trialled in a whole organism. With animals cheap to maintain, readily available, rapidly reproducing and not subject to the same strict ethical laws they have become an invaluable component of modern scientific practice.

Which of the following best illustrates the main conclusion of this argument?

A) New pharmaceuticals cannot be approved without animal experimentation.
B) Cell culture experiments are unhelpful.
C) Modern medicine would not have achieved its current standard without animal experimentation.
D) Logistically animals are easier to keep than humans for mandatory experiments.
E) All of the above.

The information below relates to questions 47 – 50:

Item Pricing

Bar chart showing cost per item (£) vs Number of Items Purchased:
- 0-99: Rugby Balls £9, Footballs £7, Tennis Balls £5
- 100-499: Rugby Balls £8, Footballs £6, Tennis Balls £4
- 500-999: Rugby Balls £7, Footballs £5, Tennis Balls £3
- 1000+: Rugby Balls £6, Footballs £4, Tennis Balls £2

The graph above shows item pricing from a wholesaler. The wholesaler is happy to deliver for a cost of £35 to companies or £5 to individuals. Any order over the cost of £100 qualifies for free delivery. Items are defined as how they come to the wholesaler therefore 1 item = 2 rugby balls or 1 football or 5 tennis balls.

Question 47:
What is the total cost to an individual purchasing 12 rugby balls and 120 tennis balls?

A) £174   B) £179   C) £208   D) £534   E) £588

Question 48:
A private gym wishes to purchase 10 of everything, how short are they of the free delivery boundary?

A) £5.00
B) £5.01
C) £10.00
D) £10.01
E) They are already over the minimum

Question 49:

What is the most number of balls that can be bought by an individual with £1,000 pounds.

A. 200
B. 250
C. 500
D. 1,000
E. 1,250

Question 50:
The wholesaler sells all his products for a profit of 120%. If he sells £1,320 worth of goods at his prices, what did he spend on acquiring them himself?

A) £400
B) £600
C) £800
D) £1,100
E) £1,120

END OF SECTION

YOU MUST ANSWER ONLY ONE OF THE FOLLOWING QUESTIONS

1) *'Nature over nurture'*. To what extent do you agree with this statement?

2) Do you think positive discrimination should be implemented in our university admission process?

3) *'Developed countries have a responsibility to help developing countries grow'*.
   Do you agree with this statement?

4) Should our education system place a greater emphasis on the sciences as opposed to the arts?

END OF TEST

Mock Paper F

Question 1:
There are four houses on a street. Lucy, Vicky, and Shannon live in adjacent houses. Shannon has a black dog named Chrissie, Lucy has a white Persian cat and Vicky has a red parrot that shouts obscenities. The owner of a four-legged pet has a blue door. Vicky has a neighbour with a red door. Either a cat or bird owner has a white door. Lucy lives opposite a green door. Vicky and Shannon are not neighbours. What colour is Lucy's door?

A) Green
B) Red
C) White
D) Blue
E) Cannot tell

Question 2:
A train driver runs a service between Cardiff and Merthyr. On average a one-way trip takes 40 minutes to drive but he requires 5 minutes to unload passengers and a further 5 minutes to pick up new ones. As the crow flies the distance between Cardiff and Merthyr is 22 miles.

Assuming he works an 8-hour shift with two 20-minute breaks, and when he arrives to work the first train is already loaded with passengers how far does he travel?

A) 132   B) 143   C) 154   D) 176   E) 198

Question 3:
The massive volume of traffic that travels down the M4 corridor regularly leads to congestion at times of commute morning and evening. A case is being made by local councils in congestion areas to introduce relief lanes thus widening the motorway in an attempt to relieve the congestion. This would involve introducing either a new 2 or 4 lanes to the motorway on average costing 1 million pound per lane per 10 miles.

Many conservationist groups are concerned as this will involve the destruction of large areas of countryside either side of the motorway. They argue that the side of a motorway is a unique habitat with many rare species residing there.

The local councils argue that with many hundreds if not thousands of cars siding idle on the motorway pumping pollutants out into the surrounding areas, it is better for the wildlife if the congestion is eased and traffic can flow through. The councils have also remarked that if congestion is eased there would be less money needed to repair the roads from car incidents with could in theory be given to the conservationist groups as a grant.

Which of the following is assumed in this passage?

A) Wildlife living on the side of the motorway cannot be re-homed.
B) Congestion causes car incidents.
C) Relief lanes have been proven to improve traffic jams.
D) A and B.
E) B and C.
F) All of the above.
G) None of the above.

Question 4:
Apples and oranges are sold in packs of 5 for the price of £1 and £1.25 respectively. Alternatively, apples can be purchased individually for 30p and oranges can be purchased individually for 50p. Helen is making a fruit salad, she remarks that her order would have cost her an extra £6.25 if she had purchased the fruit individually.

Which of the following could have been her order?
A) 15 apples 10 oranges
B) 15 apples 15 oranges
C) 25 apples 10 oranges
D) 25 apples, 15 oranges
E) 30 apples, 30 oranges

Question 5:
Janet is conducting an experiment to assess the sensitivity of a bacterial culture to a range of antibiotics. She grows the bacteria so they cover an entire petri dish and then pipettes a single drop of differing antibiotic at different locations. A schematic of her results is shown right where black represents growth of bacteria.

Which of the following best describes Janet's results?

A) This strain of bacteria is susceptible to all antibiotics used.
B) This strain of bacteria is susceptible to none of the antibiotics used.
C) E was the most effective antibiotic.
D) C was the most ineffective.
E) D is the most likely to be used in further testing.

Question 6:
Laura is blowing up balloons for a birthday party. The average volume of a balloon is 300cm^3 and Laura's maximum forced expiratory rate in a single breathe is 4.5l/min.

What is the fastest Laura could inflate 25 balloons assuming it takes her 0.5 seconds to breathe in per balloon in and somebody else ties the balloons for her?

A) 112.5 seconds
B) 122.5 seconds
C) 132.5 seconds
D) 142.5 seconds
E) 152.5 seconds

Question 7:
George reasons that A is equal to B which is not equal to C. In which case C is equal to D which is equal to E. Which of the following, if true, would most *weaken* George's argument?

A) A does not equal D.
B) B is equal to E.
C) A and B are not equal.
D) C is equal to 0.
E) None of the above

Question 8:
In a single day how many times do the hour, minute and second hands of analogue clock all point to the same number?

A) 12
B) 24
C) 36
D) 48
E) 72

Question 9:
"People who practice extreme sports should have to buy private health insurance."

Which of the following statements most strongly supports this argument?

A) Exercise is healthy and private insurance offers better reward schemes.
B) Extreme spots have a higher likelihood of injury.
C) Healthcare should be free for all.
D) People that practice extreme sports are more likely to be wealthy.

Question 10:
Explorers in the US in the 18th Century had to contest with a great variety of obstacles ranging from natural to man-made. Natural obstacles included the very nature and set up of the land, presenting explorers with the sheer size of the land mass, the lack of reliable mapping as well as the lack of paths and bridges. On a human level, challenges included the threat from outlaws and other hostile groups. Due to the nature of the settling situation, availability of medical assistance was sketchy and there was a constant threat of diseases and fatal results of injuries.

Which of the following statements is correct with regards to the above text?

A) Medical supply was good in the US in the 18th Century.
B) The land was easy to navigate.
C) There were few outlaws threatening the individual.
D) Crossing rivers could be difficult.
E) All the above.

Question 11:
The statement "The human race is not dependent on electricity" assumes what?

A) We have no other energy resource.
B) Electricity is cheap.
C) Electrical appliances dominate our lives.
D) Electricity is now the accepted energy source and is therefore the only one available.
E) All of the above.

Question 12:
Wine is sold in cases of 6 bottles. A bottle of wine holds 70cl of fluid whereas a wine glass holds 175ml. Cases of wine are currently on offer for £42 a case buy one get one free.

If Elin is hosting a 3-course dinner party for 27 of her friends, and she would like to provide everyone with a glass of wine per course, how much will the wine cost her?

A) £42   B) £84   C) £126   D) £168   E) £210

Question 13:
Hannah buys a television series in boxset. It contains a full 7 series with each series comprising 12 episodes. Rounded to the nearest 10 each episode lasts 40 minutes.

What is the shortest amount of time it could possibly take to watch all the episodes back to back?

A) 49 hours   B) 51 hours   C) 53 hours   D) 56 hours   E) 60 hours

Question 14:
Many are familiar with the story that aided in the discovery of the "germ". Semmelweis worked in a hospital where maternal death rates during labour were astronomically high. He noticed that medical students often went straight from dissection of cadavers to the maternity wards. As an experiment Semmelweis split the student cohort in half. Half did their maternity rotation instead before dissection whereas the other half maintained their traditional routine. In the new routine, maternity ward before dissection, Semmelweis recorded an enormous reduction in maternal deaths and thus the concept of the pathogen was born.

What is best exemplified in this passage?

A) Science is a process of trial and error.
B) Great discoveries come from pattern recognition.
C) Provision of healthcare is closely associated with technological advancements.
D) Experiments always require a control.
E) All of the above.

Question 15:
Jack sits at a table opposite a stranger. The stranger says here I have 3 precious jewels: a diamond, a sapphire, and an emerald. He tells Jack that if he makes a truthful statement Jack will get one of the stones, if he lies he will get nothing.

What must Jack say to ensure he gets the sapphire?

A) Tell the stranger his name.
B) Tell the stranger he must give him the sapphire.
C) Tell the stranger he wants the emerald.
D) Tell the stranger he does not want the emerald or the diamond.
E) Tell the stranger he will not give him the emerald or the diamond.

Question 16:

Simon invests 100 pounds in a saver account that awards compound interest on a 6-monthly basis at 50%. Simon's current account awards compound interest on a yearly basis at 90%.

After 2 years will Simon's investment in the saver account yield more money than it would have in the current account?

A) Yes   B) No   C) Can't tell

Question 17:
My mobile phone has a 4-number pin code using the values 1 – 9. To determine this, I use a standard algorithm of multiplying the first two numbers, subtracting the third and then dividing by the fourth. I change the code by changing the answer to this algorithm – I call this the key. What is the largest possible key?

A) 42
B) 55
C) 70
D) 80
E) 81

Question 18:
A group of scientists investigates the role of different nutrients after exercise. They set up two groups of averagely fit individuals consisting of the same number of both males and females aged 20 – 25 and weighing between 70 and 85 kilos. Each group will conduct the same 1hr exercise routine of resistance training, consisting of various weighted movements. After the workout they will receive a shake with vanilla flavour that has identical consistency and colour in all cases. Group A will receive a shake containing 50 g of protein and 50g of carbohydrates. Group B will receive a shake containing 100 g of protein and 50 g of carbohydrates. All participants have their lean body mass measured before starting the experiment.
Which of the following statements is correct?

A) The experiment compares the response of men and women to endurance training.
B) The experiment is flawed as it does not take into consideration that men and women respond differently to exercise.
C) The experiment does not consider age.
D) The experiment mainly looks at the role of protein after exercise.
E) None of the above.

Question 19:
Adam, Beth and Charlie are going on holiday together. A single room costs £60 per night, a double room costs £105 per night and a four-person room costs £215 per night. It is possible to opt out from the cleaning service and to pay £12 less each night per room.

What is the minimum amount the three friends could pay for their holiday for a three-night stay at the hotel?

A) £122   B) £144   C) £203   D) £423   E) £432

Question 20:
I have two 96ml glasses of squash. The first is comprised of $\frac{1}{6}$ squash and $\frac{5}{6}$ water. The second is comprised of $\frac{1}{4}$ water and $\frac{3}{4}$ squash. I take 48ml from the first glass and add it to glass two. I then take 72ml from glass two and add it to glass one.

How much squash is now in each glass?

A) 16ml squash in glass one and 72ml squash in glass two.
B) 40ml squash in glass one and 32ml squash in glass two.
C) 48ml squash in glass one and 32ml squash in glass two.
D) 48ml squash in glass one and 40ml squash in glass two.
E) 80ml squash in glass one and 40ml squash in glass two.

Question 21:
It may amount to millions of pounds each year of taxpayers' money; however, it is strongly advisable for the HPV vaccination in schools to remain. The vaccine, given to teenage girls, has the potential to significantly reduce cervical cancer deaths and furthermore, the vaccines will decrease the requirement for biopsies and invasive procedures related to the follow-up tests. Extensive clinical trials and continued monitoring suggest that both Gardasil and Cervarix are safe and tolerated well by recipients. Moreover, studies demonstrate that a large majority of teenage girls and their parents are in support of the vaccine.

Which of the following is the conclusion of the above argument?

A) HPV vaccines are safe and well tolerated
B) It is strongly advisable for the HPV vaccination in schools to remain
C) The HPV vaccine amounts to millions of pounds each year of taxpayers' money
D) The vaccine has the potential to significantly reduce cervical cancer deaths
E) Vaccinations are vital to disease prevention across the population

Question 22:

Anna cycles to school, which takes 30 minutes. James takes the bus, which leaves from the same place as Anna, but 6 minutes later and gets to school at the same time as Anna. It takes the bus 12 minutes to get to the post office, which is 3km away. The speed of the bus is $\frac{5}{4}$ the speed of the bike. One day Anna leaves 4 minutes late.

How far does she get before she is overtaken by the bus?

A) 1.5km   B) 2km   C) 3km   D) 4km   E) 6km

Question 23:

The set two maths teacher is trying to work out who needs to be moved up to set one and who to award a certificate at the end of term. The students must fulfil certain criteria:

Move to set one	Attendance over 95%
	Average test mark over 92
	Less than 5% homework handed in late
Awarded a Certificate	Absences below 4%
	Average test mark over 89
	At least 98% homework handed in on time

	Terry	Alex	Bahara	Lucy	Shiv
Attendance %	97	92	97	100	98
Average test mark %	89	93	94	95	86
Homework handed in on time %	96	92	100	96	98

Who would move up a set and who would receive a certificate?
A) Bahara would move up a set and receive a certificate.
B) Bahara and Lucy would move up a set and Bahara would receive a certificate.
C) Bahara, Terry and Lucy would move up a set and Bahara and Shiv would receive a certificate.
D) Lucy would move up a set and Bahara would receive a certificate.
E) Lucy would move up a set and Bahara and Terry would receive a certificate.

Question 24:

18 years ago, A was 25 years younger than B is now. In 21 years time, A will be 28 years older than B was 14 years ago. How old is A now if A is $\frac{5}{6}$B?

A) 27   B) 28   C) 35   D) 42   E) 46

Question 25:

The time now is 10.45am. I am preparing a meal for 16 guests who will arrive tomorrow for afternoon tea. I want to make 3 scones for each guest, which can be baked in batches of 6. Each batch takes 35 minutes to prepare and 25 minutes to cook in the oven and I can start the next batch while the previous batch is in the oven. I also want to make 2 cupcakes for each guest, which can be baked in batches of 8. It takes 15 minutes to prepare the mixture for each batch and 20 minutes to cook them in the oven. I will also make 3 cucumber sandwiches for each guest. 6 cucumber sandwiches take 5 minutes to prepare.

What will the time be when I finish making all the food for tomorrow?

A) 4:35pm   B) 5.55pm   C) 6:00pm   D) 6:05pm   E) 7:20pm

Question 26:

Pyramid	Base edge (m)	Volume (m³)
1	3	33
2	4	64
3	2	8
4	6	120
5	2	8
6	6	120
7	4	64

What is the difference between the height of the smallest and tallest pyramids?

A) 1m B) 5m C) 4m D) 6m E) 8m

Question 27:
The wage of Employees at Star Bakery is calculated as: £210 + (Age x 1.2) – 0.8 (100 - % attendance).
Jessica is 35 and her attendance is 96%. Samira is 65 and her attendance is 89%.

What is the difference between their wages?

A) £30.40 B) £60.50 C) £248.80 D) £263.20 E) £279.20

Question 28:
It is important that research universities demonstrate convincing support of teaching. Undergraduates comprise an overwhelming proportion of all students and universities should make an effort to cater to the requirements of the majority of their student body. After all, many of these students may choose to pursue a path involving research and a strong education would provide students with skills equipped towards a career in research.

What is the conclusion of the above argument?
A) Undergraduates comprise an overwhelming proportion of all students.
B) A strong education would provide a strong foundation and skills equipped towards a career in research.
C) Research universities should strongly support teaching.
D) Institutions should provide undergraduates with a high-quality learning experience.
E) Research has a greater impact than teaching and limited universities funds should mainly be invested in research.

Question 29:
American football has reached a level of violence that puts its players at too high a level of risk. It has been suggested that the NFL, the governing body for American football should get rid of the iconic helmets. The hard-plastic helmets all have to meet minimum impact-resistance standards intended to enhance safety, however in reality they gave players a false sense of security that only resulted in harder collisions. Some players now suffer from early onset dementia, mood swings and depression. The proposal to ban helmets for good should be supported. Moreover, it would prevent costly legal settlements involving the NFL and ex-players suffering from head trauma.

What is the conclusion of the above argument?
A) Sports players should not be exposed to unnecessary danger.
B) Helmets give players a false sense of security.
C) Players can suffer from early onset dementia, mood swings and depression.
D) The proposal to ban helmets should be supported.
E) American football is too violent and puts its players at risk.

Question 30:
At the final stop (stop 6), 10 people get off the tube. At the previous stop (stop 5) $\frac{1}{2}$ of passengers got off. At stop 4, $\frac{3}{5}$ of passengers got off. At stop 3, $\frac{1}{3}$ of passengers got off and at stops 1 and 2, $\frac{1}{6}$ of passengers got off.

How many passengers got on at the first stop?

A) 10   B) 36   C) 90   D) 108   E) 3600

Question 31:
Everyone likes English. Some students born in spring like Maths and some like Biology. All students born in winter like Music and some like Art. Of those born in autumn, no one likes Biology, and everyone likes Art.

Which of the following is true?

A) Some students born in spring like both Biology and Maths.
B) Students born in spring, winter, and autumn all like Art.
C) No one born in winter or autumn likes Biology.
D) No one who likes Biology also likes Art.
E) Some students born in winter like 3 subjects.

Question 32:
Until the twentieth century, the whole purpose of art was to create beautiful, flawless works. Artists attained a level of skill and craft that took decades to perfect and could not be mirrored by those who had not taken great pains to master it. The serenity and beauty produced from movements such as impressionism has however culminated in repulsive and horrific displays of rotting carcasses designed to provoke an emotional response rather than admiration. These works cannot be described as beautiful by either the public or art critics. While these works may be engaging on an intellectual or academic level, they no longer constitute art.

Which of the following is an assumption of the above argument?
A) Beauty is a defining property of art.
B) All modern art is ugly.
C) Twenty first century artists do not study for decades.
D) The impressionist movement created beautiful works of art.
E) Some modern art provokes an emotional response.

Question 33:
The cost of sunglasses is reduced over the bank holiday weekend. On Saturday, the price of the sunglasses on Friday is reduced by 10%. On Sunday the price of the sunglasses on Saturday is reduced by 10%. On Monday, the price of the sunglasses on Sunday is reduced by a further 10%. What percentage of the price on Friday is the price of the sunglasses on Monday?

A) 55.12%   B) 59.10%   C) 63.80%   D) 70.34%   E) 72.9%

Question 34:
Putting the digit 7 on the right-hand side of a two-digit number causes the number to increase by 565. What is the value of the two-digit number?

A) 27
B) 52
C) 62
D) 66
E) 627

Question 35:

When folded, which box can be made from the net shown below?

Question 36:

The grid is comprised of 49 squares. The shape's area is 588cm². What is its perimeter in cm?

A) 26   B) 49   C) 84   D) 126   E) 182

Questions 37-39 refer to the following information:

$$BMI = weight\ (kg) \div height^2\ (m^2)$$

Men	BMR= (10 x weight in kg) + (6.25 x height in cm) – (5 x age in years) + 5
Women	BMR= (10 x weight in kg) + (6.25 x height in cm) – (5 x age in years) -161

Recommended Intake:

Amount of Exercise	Daily Kilocalories required
Little to no exercise	BMR x 1.2
Light exercise 1-3 days per week	BMR x 1.375
Moderate exercise 3-5 days per week	BMR x 1.55
Heavy exercise 6-7 days per week	BMR x 1.725
Very heavy exercise twice per day	BMR x 1.9

Question 37:
A child weighs 35kg and is 120cm tall. What is the BMI of the child to the nearest 2 decimal places?

A) 0.0024   B) 24.28   C) 24.31   D) 42.01   E) 42.33

Question 38:
What is the BMR of a 32-year-old woman weighing 80kg and measuring 1.7m in height?

A) 643.7 kcal   B) 1537 kcal   C) 1541.5 kcal   D) 1707.5 kcal   E) 2707.5 kcal

Question 39:
What is the recommended intake of a 45-year-old man weighing 80kg and measuring 1.7m in height who does little to no exercise each week?

A) 1642.5 kcal   B) 1771.8 kcal   C) 1851 kcal   D) 1971 kcal   E) 2712.5 kcal

Questions 40 and 41 relate to the following passage:
The achievement levels of teenagers could be higher if school started later. Teenagers are getting too little sleep because they attend schools that start at 8:30am or earlier. A low level of sleep disrupts the body's circadian rhythms and can contribute to health problems such as obesity and depression. Some doctors are now urging schools to start later in order for teenagers to get sufficient sleep; ideally 8.5 to 9.5 hours each night. During adolescence, the hormone melatonin is released comparatively later in the day and the secretion levels climb at night. Consequently, teenagers can have trouble getting to sleep earlier in the night before sufficient melatonin is present. A school in America tested this idea by starting one hour later and the percentage of GCSEs at grades A* - C increased by 16%. Schools in the UK should follow by example and shift start times later.

Question 40:
Which of the following is a flaw in the above argument?

A) Slippery slope.
B) Hasty generalisation.
C) It confuses correlation with cause.
D) Circular argument.
E) Schools should not prioritise academic achievement.

Question 41:
Which of the following, if true, would most strengthen the above argument?

A) Teenagers who are more alert will disrupt the class more.
B) Getting more sleep at night is proven to increase activity levels.
C) American schools and British schools teach the same curriculum.
D) The school in America did not alter any other aspects of the school during the trial for example curriculum, teachers, and number of students in each class.
E) Teachers get tired towards the end of the school day and are less effective.

Question 42:
The UK energy market is highly competitive. In an effort to attract more business and increase revenue, the company EnergyFirst has invested significant funds into its publicity sector. Last month, they doubled their advertising expenditures, becoming the energy company to invest the greatest proportion of investment into advertising. As a result, it is expected that EnergyFirst will expand its customer base at a rate exceeding its competitors in the ensuing months. Other energy companies are likely to follow by example.

Which of the following, if true, is most likely to weaken the above argument?

A) Other companies invest more money into good customer service.
B) Research into the energy industry demonstrates a low correlation between advertising investment and new customers.
C) The UK energy market is not highly competitive.
D) EnergyFirst currently has the smallest customer base.
E) Visual advertising heavily influences customers.

Question 43:
The consumption of large quantities of red meat is suggested to have negative health ramifications. Carnitine is a compound present in red meat and a link has been discovered between carnitine and the development of atherosclerosis, involving the hardening, and narrowing of arteries. Intestinal bacteria convert carnitine to trimethylamine-N-oxide, which has properties that are damaging to the heart. Moreover, red meat consumption has been associated with a reduced life expectancy. It may be that charring meat generates toxins that elevate the chance of developing stomach cancer. If people want to be healthy, a vegetarian diet is preferable to a diet including meat. Vegetarians often have lower cholesterol and blood pressure and a reduced risk of heart disease.

Which of the following is an assumption of the above argument?

A) Diet is essential to health and we should all want to be healthy.
B) Vegetarians do the same amount of exercise as meat eaters.
C) Meat has no health benefits.
D) People who eat red meat die earlier.
E) Red meat is the best source of iron.

Question 44:
Auckland is 11 hours ahead of London. Calgary is 7 hours behind London. Boston is 5 hours behind London. The flight from Auckland to London is 22 hours, but the plane must stop for 2 hours in Hong Kong. The flight from London to Calgary is 8 hours 30 minutes. The flight from Calgary to Boston is 6 hours 30 minutes. Sam leaves Auckland at 10am for London. On arrival to London, he waits 3 hours then gets the plane to Calgary. Once in Calgary, he waits 1.5 hours and gets the plane to Boston. What time is it when Sam arrives in Boston?

A) 13:30pm   B) 22.30pm   C) 01:00am   D) 01:30am   E) 03:30am

Question 45:
Light A flashes every 18 seconds, light B flashes every 33 seconds and light C flashes every 27 seconds. The three lights all flashed at the same time 5 minutes ago.

How long will it be until they next all flash simultaneously?

A) 33 seconds B) 294 seconds C) 300 seconds D) 333 seconds E) 594 seconds

Question 46:
There are 30 students in a class. They must all play at least 1 instrument, but no more than 3 instruments. 70% play the piano, 40% play the violin, 20% play the guitar and 10% play the saxophone. Which of the following statements must be true?

1. 3 or more students play piano and violin.
2. 12 students or less play the piano and the violin.
3. 9 or more students do not play piano or violin.

A) 1 only B) 1 and 2 C) 2 only D) 2 and 3 E) 3 only

Question 47:
Neil, Simon and Lucy are playing a game to see who can role the highest number with two dice. They start with £50 each. The losers must halve their money and give it to the winner of each game. If it is a draw, the two winners share the loser's money. If all three tie, then they keep their money. Neil wins game 1, Simon and Lucy win game 2 and 5 and Lucy wins games 3 and 4.

How much money does Lucy gain?

A) £15.63 B) £75.00 C) £75.16 D) £78.13 E) £128.10

Question 48:
Drivers in the age group 17-19 comprise 1.5% of all drivers; however, 12% of all collisions involve young drivers in this age category. The RAC Foundation wants a graduated licensing system with a 1-year probationary period with restrictions on what new drivers can do on roads. Additionally, driving instructors need to emphasise the dangers of driving too fast and driving tests should be designed to make new drivers more focused on noticing potential hazards. These changes are essential and could stop 4,500 injuries on an annual basis.

What is the assumption of the above argument?
A) Young drivers are more likely to have more passengers than other age groups.
B) Young drivers spend more hours driving than older drivers.
C) Young drivers are responsible for the collisions.
D) The cars that young people drive are unsafe.
E) Most young drivers involved in accidents are male.

Question 49:
The mean weight of 6 apples is 180g. The lightest apple weighs 167g. What is the highest possible weight of the heaviest apple?

A) 193g B) 225g C) 235g D) 245g E) 255g

Question 50:
"Sugar should be taxed like alcohol and cigarettes." Which of the following arguments most supports this claim?

A) Sugar can cause diabetes.
B) Sugar has high addictive potential and is associated with various health concerns.
C) High sugar diets increase obesity.
D) People that eat a lot of sugar are more likely to start abusing alcohol.
E) None of the above.

END OF SECTION

YOU MUST ANSWER ONLY ONE OF THE FOLLOWING QUESTIONS

1   *'Social media has risen and fallen'.*
          Discuss.

2   Do you think the right to press freedom has become increasingly dangerous and polarising to society?

3   *'Artificial intelligence replaces humans and leads to unemployment'.*
          Do you agree with this statement?

4   Does our society place too much emphasis on a university education?

END OF TEST

ANSWERS

Answer Key

Paper D	Paper E	Paper F
1 C	1 C	1 B
2 D	2 D	2 E
3 B	3 B	3 E
4 E	4 D	4 D
5 A	5 D	5 E
6 C	6 B	6 A
7 B	7 A	7 B
8 B	8 D	8 B
9 D	9 C	9 B
10 A	10 E	10 D
11 B	11 D	11 C
12 E	12 E	12 B
13 D	13 D	13 E
14 C	14 E	14 E
15 D	15 A	15 C
16 A	16 D	16 A
17 B	17 B	17 D
18 E	18 C	18 D
19 C	19 B	19 D
20 E	20 D	20 D
21 B	21 E	21 B
22 C	22 D	22 B
23 E	23 E	23 B
24 A	24 B	24 C
25 D	25 C	25 D
26 D	26 C	26 D
27 C	27 B	27 A
28 D	28 B	28 C
29 E	29 C	29 D
30 E	30 C	30 D
31 B	31 B	31 E
32 C	32 E	32 A
33 D	33 D	33 E
34 C	34 A	34 C
35 D	35 B	35 E
36 C	36 D	36 E
37 E	37 A	37 E
38 D	38 E	38 C
39 E	39 D	39 D
40 B	40 B	40 B
41 E	41 D	41 D
42 D	42 E	42 B
43 D	43 D	43 A
44 D	44 D	44 A
45 B	45 B	45 B
46 A	46 E	46 B
47 E	47 D	47 D
48 C	48 C	48 C
49 E	49 B	49 D
50 E	50 A	50 B

Raw to Scaled Scores

1	10.5	11	38.5	21	49.5	31	59.0	41	70.5
2	18.0	12	40.0	22	50.5	32	60.0	42	72.0
3	22.5	13	41.0	23	51.5	33	61.0	43	74.0
4	25.5	14	42.0	24	52.5	34	62.0	44	75.5
5	28.5	15	43.5	25	53.5	35	63.0	45	78.0
6	30.5	16	44.5	26	54.5	36	64.0	46	80.5
7	32.5	17	45.5	27	55.0	37	65.0	47	83.5
8	34.0	18	46.5	28	56.0	38	66.5	48	88.0
9	35.5	19	47.5	29	57.0	39	67.5	49	95.5
10	37.0	20	48.5	30	58.0	40	69.0	50	103

Mock Paper Answers

Mock Paper D: Section 1

Question 1: C

Answer B) is completely irrelevant to what the manager is saying, so is incorrect. A) and E) are also incorrect as the manager is simply talking about ticket sales. He has not mentioned anything about the relevant popularity of folk music, or how much the band should value profit. D) is incorrect as the manager is simply saying that the band will have higher ticket sales in France than in Germany, so other countries are not relevant.

C) is correct as Germany could still have higher ticket sales for folk music than France despite the recent changes in ticket sales.

Question 2: D

E) is completely irrelevant to Rob's conclusion. C) is also irrelevant as Rob has only concluded that Tom is unable to take part, and has said nothing about his own eligibility. B) is incorrect as whether Tom intended to take part in the mountain climbing activities does not affect whether he is able to under the clause Rob has found.

Meanwhile, A) is explicitly stated in the question, as the clause applies to holidays organised by Wild Africa Inc. and Tom and Rob's holiday is organised by Wild Africa Inc. Therefore it is not an assumption.

D) is correct as at no point has it been stated that Tom suffers from *severe* asthma. It has only been stated that Tom suffers from asthma, nothing has been said about the severity. If Tom does not suffer from severe asthma, then Rob's conclusion is incorrect.

Question 3: B

Only B) is not an assumption, as it is stated in the question that both Grace and Rose departed at 5:15pm. The other answers are all assumptions. At no point has it been stated that both the girls are walking, or that they will walk at the same speed. If either of these points are incorrect, we cannot definitely state that they will arrive home at the same time. Therefore A) and E) are assumptions. Also, it has not been stated that the gymnastics class is being held at the local gymnasium. If this is not the case, then we cannot know how far Grace and Rose have to walk, and therefore cannot state that they will arrive home at the same time. Therefore, C) is an assumption. Equally, if Grace gets lost, she may arrive home after Rose, so D) is an assumption.

Question 4: E

Answer A) is completely irrelevant to John's conclusions, as the speed of travel has no effect on the train's destination. D) is also irrelevant as other destinations from King's Cross station also bear no effect on John's conclusion. Meanwhile, B) is incorrect as John's conclusions refer to travelling to Edinburgh by train, so the possibility of travelling by aeroplane has no effect. C) is not an assumption because John's conclusion is in the present tense, referring to journeys made at the moment, so future developments have no effect.

E) is an assumption John has made. Only two other stations in London have been mentioned. At no point has it been mentioned that there are no other stations in London that John could travel from.

Question 5: A

B), C) and D) are all stated in the question, and so are not assumptions. D) is stated as the question states that safety features cause people to drive faster, and then states that higher driving speeds are responsible for many high-speed crashes. E) directly contradicts something stated in the question (the question actually states that high-speed crashes are causing thousands of deaths each year). Also, E) bears no influence on the conclusion of the argument. Therefore, E) is not an assumption.

A) is an assumption because it is required to be true for the argument's conclusion to be valid, but at no point is it stated that a spike would cause people to drive slower.

Question 6: C

B) and D) are both stated in the question. A) is also stated as the question states that Tanks were a hugely influential factor in ALL battles in World War 2.

E) is not stated but is not an assumption as it is not required to be true for the argument's conclusion to be valid.

C) However, is required to be true for the conclusion to be valid and yet is never stated in the question, so it is an assumption.

Question 7: B

D) is irrelevant to the argument's conclusion, whilst A) and E) are also irrelevant as the argument does not directly imply either of these things (and even if it did they are irrelevant to the argument's conclusions so are not flaws).

C) is incorrect because the argument states that the Prussian arrival was essential to the British victory, so C) is not an assumption.

B), however, is never stated in the question, but is needed to be true for the argument's conclusion to be valid.

Question 8: B

D) and E) are both entirely irrelevant to waiting times, so are not flaws.

C) is not correct, as the question states that busier ports have longer queuing times. A) is also incorrect as the question states that Bordeux is the busiest port in France, so Calais is definitely less busy than Bordeux. Therefore, Porto cannot be busier than Bordeux but less busy than Calais.

B) is a flaw, as the fact that Bilbao was busiest last year does not necessarily mean it will be the busiest this year.

Question 9: D

A) is irrelevant to the argument's conclusion, and therefore not a flaw. C) is also irrelevant as other problems stemming from inhalers do not change the fact that inhalers could prevent many hospital trips.

B) and E) are incorrect. The argument only states that inhalers could prevent many hospital trips. It does not assume that inhalers could prevent asthma attacks, and does not state that all the hospital trips could be prevented.

D) is a valid flaw with the argument. At no point is it stated that inhalers stocked in school could be quickly available to children, and yet the argument states that hospital trips could be prevented if inhalers were *quickly available*.

MOCK PAPER D SECTION 1 — ANSWERS

Question 10: A

At no point is A) stated, but if aeroplanes are not a major source of carbon dioxide then it does not follow that they are largely responsible for the damage caused by global warming. Therefore A) is merely an assumption.
B) and C) are both stated in the question, whilst D) is irrelevant to the conclusion. E), meanwhile, is stated, as the question states that *we must now seek to curb air traffic in order to save the world's remaining natural environments.*

Question 11: B

A) and C) can be inferred, as the question states that these things would happen. Meanwhile, D) and E) actually serve to reinforce the argument's conclusion that the research into a new cure will not be successful. Therefore, they are not flaws in the argument's reasoning.
The point raised by B) does weaken the argument, and is a valid flaw in the argument's reasoning.

Question 12: E

A), C) and D) are all irrelevant to the argument's main conclusion, namely that Egypt was a powerful nation and must therefore have had a very strong military.
B) is a conclusion from the argument, but goes on to support E). If a nation required a very strong military to be a powerful nation, then it follows that if Egypt was a powerful nation it must have had a very strong military. Therefore, B) is an intermediate conclusion within the argument. E) is the *main* conclusion of the argument.

Question 13: D

At no point does the argument state or imply that we should not be concerned about damage to the polar ice caps, or that reducing energy consumption will not reduce CO2 emissions. Therefore, B) and E) are incorrect.
C) could be described as an assumption made in the argument, and is therefore not a conclusion.
A) goes beyond what the argument says. The argument does not say there are no environmental benefits to reducing energy consumption; it merely says it will not help the polar ice caps. Therefore A) is incorrect and C) is a valid conclusion from the argument.

Question 14: C

A), B) and D) are all in direct contradiction to statements made in the passage, so cannot be conclusions. E), meanwhile, does not contradict the argument, but at no point does the argument say that the dangerous isotope was not effective at relieving nausea, so E) is not a conclusion.
However, the fact that the company followed the required level of testing and still did not detect the dangerous isotope does suggest that the required level of testing was not sufficient to identify Isotopes, so C) is correct.

Question 15: D

A) and E) are both intermediate conclusions which go on to support the main conclusion, which is that given in D).
B) is irrelevant, and is also not stated or implied in the argument, whilst C) is an assumption made in the argument, and is therefore not a conclusion.

Question 16: A

At no point is it stated or implied that car companies should prioritise profits over the environment, so C) is incorrect. Neither is it stated that the public do not care about helping the environment, so E) is incorrect.
B) is a reason given in the argument, whilst D) is impossible if we accept the argument's reasons as true, so neither of these are conclusions.
The conclusion is that given in A).

Question 17: B

E) is contradictory to the main conclusion of the argument.
A), C) and D) are all reasons which go on to support the main conclusion of the argument, which is given in B). If we accept A), C) and D) as true, then it follows readily that the statement given in B) is true. Therefore, B) is the main conclusion.

MOCK PAPER D SECTION 1 — ANSWERS

Question 18: E

A) is contradictory to the researcher's argument, whilst D) is a completely irrelevant suggestion.

The researcher makes no suggestion that a decrease in productivity will result from not organising a staff day out, so C) is incorrect.

The researcher accepts that there will be a short-term decrease in productivity from organising a staff day out, but that it will be made up for by the long-term increase in productivity. Therefore B) is incorrect, as the researcher is not suggesting the short term decrease is ignores. E) is more in line with the researcher's argument, so E) is the main conclusion.

Question 19: C

B) and D) are irrelevant suggestions that can be safely ignored. E) is in direct contradiction to a statement in the passage that the 2 candidates were clearly more suitable than the others.

A) is contradictory to what the manager has said, whilst C) follows on from what the manager says. Therefore, C) is a conclusion which is much more readily drawn from the passage.

Question 20: E

The passage is suggesting a system should be used in which people pay for *whichever* services they use. The new system is not referring to *how much* a customer uses a service, simply *which* services are used.

All of the answers follow this reasoning except for E). E) refers to a system where people pay according to *how much* they use a service, rather than paying for whichever services they use. Therefore, E) follows a different line of reasoning from the passage.

Question 21: B

B) is an underlying assumption in the Transport Minister's argument. If rural areas have plenty of passengers, her assertion that rail companies will not run many services to these areas does not follow from her reasoning. Therefore, if B) is true, it strengthens the transport minister's argument.

Meanwhile, D) would actually weaken the transport minister's argument, suggesting that privatisation would not lead to less service for rural areas.

C) is irrelevant as the transport minister is arguing about how rural communities will be cut off by a privatised system. She is not referring to the quality or price of rail services under a publically subsidised system.

A) and E) are completely irrelevant points, which have no effect at all on the strength of the Transport Minister's argument.

Question 22: C

The Argument follows a pattern of "**IF** A happens, B **WILL** happen. A Happens, therefore B **WILL** happen". C) is the only answer which follows this pattern of reasoning.

A) reasons as "**IF** A happens, B **WILL** happen", but then incorrectly concludes that "If B is to happen, A **must** happen". This is incorrect. In this example, there may be other cheaper airports to fly from apart from Gatwick.

D) is also incorrect, stating that "A **must** happen for B to happen" but then concluding "A happens, therefore B **will** happen". This is incorrect, as the fact that A is needed for B to happen offers no guarantee that B will happen if A does.

B) and E) both follow correct reasoning, but different to that in the question. B) and E) reason as "A **must** happen for B to happen. B happens, therefore A **must** have happened". This is not the same as saying "If A happens, B **will** happen", as in the question.

Question 23: E

The passage follows principle that everybody should pay an amount depending on how wealthy they are, which is then used to fund services that are allocated based on need.

B), C) and E) both refer to services being allocated based on need, not ability to pay. However, C) refers to companies responsible for the problems being made to pay for the service, and B) refers to everybody paying a fixed amount to fund the service. E) refers to people paying an amount dependent on personal wealth. Therefore E) best illustrates the principle of the passage.

The other answers do not refer to a service being allocated based on need. D) directly contradicts this principle, whilst A) is not referring to a service that is being funded/provided.

MOCK PAPER D SECTION 1 — ANSWERS

Question 24: A
B) and E) are irrelevant points which do not affect the strength of Lucy's argument.
C) and D) would both serve to strengthen Lucy's argument. C) suggests that running costs will be low, whilst D) suggests that visitor centres are profitable. Both of these, if true, serve to suggest that opening up visitor centres will be profitable for the park, therefore supporting Lucy's argument.
A), however, would weaken Lucy's argument by suggesting that visitor centres will not be profitable.

Question 25: D
The principle used in the argument is that there is no point in engaging in debate with people who are wrong about the facts on a particular issue. Only D) follows this line of reasoning.
B) and E) are both in direct contradiction to this principle, whilst A) is an irrelevant point, not referring to who we should engage in debate with.
C) refers to who Kerry should talk to, but does not explicitly state that she should not talk to those who are wrong about the issue. Therefore C) is not as effective an illustration of the principle at hand as D).

Question 26: D
If the announcement is accurate to the nearest 10 minutes, this means that the soonest the train will arrive in London is 115 minutes after the announcement, which is 17:25. The final destination is 10 minutes from King's Cross, so the earliest time I might arrive there is 17:35.

Question 27: C
When the sister's son is home, she has to buy a carton of washing powder 1.2 times as often, so she must be doing 1.2 times as many loads of washing. If we let x be the number of people living at home when the son is home, the number of loads of washing when he is home is 2+x, whereas the number when is not at home is 2+x-1=1+x. Therefore 2+x must equal 1.2(1+x). Rearranging this we get 0.8=0.2x, so x is 0.8 divided by 0.2, which is 4. So 4 people are living in the house when the son is home.

Question 28: D
The total time each train can run is 18 hours a day. Each journey takes the train 30 minutes (24+4+2). So each train can make 36 journeys a day. Therefore the total journeys made by the shuttle service per day will be 2x36 (because there are 2 trains) so the answer is 72.

Question 29: E
The lowest common multiple of 6, 4 and 2.5 is 60. Hence trains from all 3 lines will arrive at the same time every 60 minutes. If the last time they did was 4 minutes ago, it will hence be 60-4=56 minutes until they do so again. Therefore the answer is E.

Question 30: E
If Sam buys the invitations, she will spend £90 (90 x £1 each).
If Sam makes the invitations, she will need enough supplies for 94 invitations, which will be:
➢ 4 packs of red paper at £2 each = £8
➢ 7 rolls of ribbon at £3 each = £21
➢ 4 packs of gold stickers at £1 each = £4
➢ 1 stamper = £8
➢ 2 ink pads at £4 each = £8
➢ 5 packs of cream card at £2 each = £10

Adding these up, we get that the total spent on making the invitations herself is £59. Therefore she saves £90-£59=£31 by making them rather than buying them. Therefore the answer is E.

MOCK PAPER D SECTION 1 — ANSWERS

Question 31: B
If there are at least as many boys as girls in the class of 36, then there are at least 18 boys and at least 9 boys have brown eyes. If two thirds of the class have brown hair and at least as many boys as girls have brown hair, at least two thirds of the boys in the class have brown hair, so at least 12 boys have brown hair. There are only 18 boys in the class, so of the 9 boys who have brown eyes and 12 who have brown hair, at least 3 of these must be the same boys. So at least 3 boys in the class have both brown hair and brown eyes.

Question 32: C
The total amount of dilute squash needed is (300ml x 8) + (400ml x 3) = 3600ml. Mandy has 600ml of concentrated squash so she needs 3000ml of water to make up the right amount. There should be 3000ml:600ml of water:concentrated squash so the ratio needed is 5:1

Question 33: D
If we assumed that none of the options in any block were the same, there would be 4x4x4x4 = 256 different sets of options. However, Psychology and Mathematics are in 2 different blocks so there will be some options that cannot be taken together. There are 16 sets of 4 options that involve Mathematics twice (Mathematics in Blocks A and C and then any combination of options in the other blocks) and 16 that involve Psychology twice (Psychology in Blocks A and B and any combination of options in the other blocks), so we need to take off 32 from the total options we calculated. Hence the total number of sets of possible options is 256-32=224.

Question 34: C
We need to work out both what the cost of living will be in 4 years time and how much savings there will be in 4 years time.
The cost of living rises by 3% a year, so using the table we can see that at the end of Year 1 the cost of living is £19,570.00. Then at the end of Year 2, the cost of living is £20,157.10. At the end of Year 3, the cost of living is £20,761.81. At the end of Year 4, the cost of living is £21,384.67.
The savings rise by 4% each year until they exceed £20,000, so after Year 1 they are £19,760.00. After Year 2, they are £20,550.40. This means that for the last 2 years, they will yield 5%. So after Year 3, they will be £21,577.92, and after Year 4, they will be £22,656.82.
We can then work out the difference between these two amounts. We only need an approximation, so we can do £22,656 - £21,384 = £1272, which to the nearest £100 is £1300. Hence the answer is C.

Question 35: D
Amaia needs to arrive by 1100 so the latest bus she can catch is the one which arrives at Northtown University at 1053. This arrives at Northtown West station at 1019, or Northtown Central station at 1040. The latest train she can catch to make this bus is the one that gets to Northtown Central for 1028, and leaves Southtown at 0943. Amaia lives a 10 minute walk from Southtown station so she will need to leave her house at 0933 to get to the interview on time. Therefore the answer is D.

Question 36: C
The instructions say that Shaniqua plays in the square which will stop Summer being able to win straight away, so Shaniqua must play in 4. Summer then needs to play in a square where there will be 2 different options to make a line on the turn afterwards, so that Shaniqua cannot block both of them. If Summer plays in 1, she can make a line by playing in either 5 or 6 the next turn, so Shaniqua cannot stop her winning. If Summer plays in 2, she cannot make a line on the next turn at all. If Summer plays in 3, she can only make a line by playing in 6 the next turn and so Shaniqua can stop her. If Summer plays in 5, she can only make a line by playing in 5 the next turn and so Shaniqua can stop her. If Summer plays in 6, she can make a line by playing in either 1 or 3 the next turn, so Shaniqua cannot stop her winning. Therefore she either needs to play in 1 or 6 to be able to be certain of winning the next time. Therefore the answer is C.

Question 37: E
There are 16 squares of dimension 1. There are 9 squares of dimension 2 (one in each corner, one halfway across each side and one right in the middle). There are 4 squares of dimension 3 (one in each corner). There is 1 square of dimension 4. Therefore the total number of squares is 16+9+4+1=30. Therefore the answer is E.

MOCK PAPER D SECTION 1 — ANSWERS

Question 38: D
Any date in October 2014 will have a 1 in both the month and year so we have to go back to September. 30th September will have a 0 in both the date and month. 29th September will have a 9 in both the date and the month. Therefore the last date with every number different in the 6 digit format will be 28-09-2014, which is 14 days before the 12th October. Therefore the answer is D.

Question 39: E
Ashley has to be sat in the front left seat so there are only two seats left in the front row. Bella and Caitlin have to be sat in different rows, so one of them must be sat in the front row and one in the back row. Now there is only one seat left in the front row, so there is not room for Danielle and her teaching assistant to both sit there. Therefore Danielle and the teaching assistant must take the two remaining seats in the back row. Therefore Emily must sit on the front row as there are no seats remaining in the back row. Emily cannot sit in the middle seat due to her mobility issues, so she must sit in the front right seat. Therefore the answer is E.

Question 40: B
- The volume of the box with 10cm squares cut out is $10*100*100 = 100000cm^3$
- The volume of the box with 20cm squares cut out is $20*80*80 = 128000cm^3$
- The volume of the box with 30cm squares cut out is $30*60*60 = 108000cm^3$
- The volume of the box with 40cm squares cut out is $40*40*40 = 64000cm^3$
- The volume of the box with 50cm squares cut out is $50*20*20 = 20000cm^3$

Therefore the biggest box is the one with the 20cm squares cut out, so the answer is B.

Question 41: E
We can calculate all the rental yields as follows:
- House A: (700x12)/168000 = 0.05
- House B: (40x125x4)/200000 = 20000/200000 = 0.10
- House C: (600x12)/144000 = 7200/144000 = 0.05
- House D: (2000x12)/240000 = 24000/240000 = 0.10

House E: (200x52)/100000 = 10400/100000. We can see by observation that this is > 0.1 as 10000/100000 would equal 0.1, therefore there is no need to work this out to be able to say that this is the house with the highest yield.

Question 42: D
Let the number of invitations with the extra information in be m. Invitations with extra information in cost £0.70 to send and invitations without cost £0.60. Therefore the total cost of posting is £0.70m + £0.60(50-m) and this is equal to £33. 33=0.70m-0.60m+30. 3=0.1m therefore m=30. So the number of invitations with extra information in is 30. Therefore the answer is D.

Question 43: D
If the median is 6, the 3rd number when the numbers are written in order is 6. If the mode is 4, there must be at least two 4s (and can only be two 4s, because there are only 2 numbers less than 6 due to what we know about the median). Therefore the smallest three numbers in the set are 4, 4, 6. For the mean to be 8, the numbers must add up to 5 times 8 = 40. Therefore the largest two numbers must add up to 40-(4+4+6)=26. Therefore the answer is D.

Question 44: D
The hospital director's comments make it abundantly clear that the most important aspect of the new candidate is good surgery skills, because the hospital's surgery success record requires improvement. If we accept his reasoning as being true, then it is clear that the candidate who is most proficient at surgery should be hired, and patient interaction should not be the deciding factor. Thus, Candidate 3 should be hired, as suggested by D.

Question 45: B

We can calculate the prices if the child's fare is a half of the adult fare, then as if the child's fare is three quarters of the adult fare, and choose the multiple of 20 that lies between the two.

If the child's fare is half the adult fare, then £4.20 is twice the adult fare as this is the price for one adult and two children each of whom pay half the adult price, so it is the equivalent of paying for 2 adults. Hence the adult fare is £2.10 and the child fare is £1.05.

If the child's fare is three quarters of the adult fare, then £4.20 is two and a half times the adult fare. Hence the adult fare is £1.68 and the child fare is 84p.

So we know the child fare is between 84p and £1.05. We also know it is a multiple of 20p. Hence the child fare is £1. The adult fare is between £1.68 and £2.10, so on the basis that it is a multiple of 20p it could either be £1.80 or £2. However, we know the child fare is more than half the adult fare, so it cannot be £2 as this would mean the child fare was half of the adult fare. Therefore the adult fare must be £1.80 and the difference between the two is £1.80-£1 = 80p. Therefore the answer is B.

Question 46: A

Each group involves 10 matches (each team plays each other team once, so if the teams are A through to E, the 10 matches are A-B, A-C, A-D, A-E, B-C, B-D, B-E, C-D, C-E, D-E). There are 8 groups, so in total in the group stage there are 80 matches. 16 teams then progress from the group stage, meaning 8 matches in this round. There are then 4 matches in the next round, 2 matches in the round after and then is the final. Therefore in total there are 80+8+4+2+1 matches, = 95. Therefore the answer is A.

Question 47: E

We can work out the code for each number and see which one equals 3.
- The code for A is (3x4) = 12, divided by 6 = 2, minus 1 = 1
- The code for B is (9x8) = 72, divided by 6 = 12, minus 4 = 8
- The code for C is (5x4) = 20, divided by 2 = 10, minus 3 = 7
- The code for D is (7x8) = 56, divided by 4 = 14, minus 8 = 6
- The code for E is (6x8) = 68, divided by 4 = 17, minus 9 = 3

Therefore the pin number with the code 3 is E, 6849.

Question 48: C

If it will cost Niall £2 more to pay per session than to buy membership, the combination of classes and gym sessions he is going to attend must cost £32. The only one of these combinations which costs £32 is C (5 x gym sessions at £4, 6 x classes at £2). Hence the answer is C.

Question 49: E

If I can lift boxes up to half my weight, the total I can lift at a time is 30kg. If the empty box weighs 0.5kg and each book is 2kg, up to 14 books can be lifted per box (so each full box weighs 28.5kg). 120 divided by 14 is between 8 and 9. Hence it will take 9 trips to move all the books.

Question 50: E

- At the end of the first year, Jenny gets a rise of 3%, which is £720, rounded up to the nearest £1000, which is £1000. Hence her salary at the end of the first year is £25000.
- At the end of the second year, Jenny gets a rise of 6%, which is £1500, rounded up to the nearest £1000, which is £2000. Hence her salary at the end of the second year is £27000.
- At the end of the third year, Jenny gets a rise of 9%, which is £2430, rounded up to the nearest £1000, which is £3000. Hence her salary at the end of the third year is £30000.
- At the end of the fourth year, Jenny gets a rise of 12%, which is £3600, rounded up to the nearest £1000, which is £4000. Hence her salary at the end of the fourth year is £34000.
- At the end of the fifth year, Jenny gets a rise of 15%, which is £5100, rounded up to the nearest £1000, which is £6000. Hence her salary at the end of the fifth year (and hence after 5 and a half years) is £40000. Hence the answer is E.

END OF SECTION

Mock Paper D: Section 2

1. Describe a scientific model of a human brain. Justify any materials you might need, how you would connect the separate components and highlight any limitations your model might have.

This is a very broad question and can be taken in many different ways – there is no correct model. The essay should be structured as follows:

- Introduction – general overview of the model, describing the important elements of the human brain and how they could be portrayed into a model. The introduction may mention the limitations of any model of a brain, in that we do not understand the human brain fully and it is far too complex to be modelled by anything other than itself.
- Main body – organized into paragraphs describing each component. All models should follow a logical structure, with clear justification for each decision made. There should be a continual reference to the human brain in the justification and clear links between paragraphs describing how the model could be achieved.
- Conclusion – this should summarize the main components of the model. Here it would be relevant to mention any major limitations, or any possible developments that could be made to the model in the future. The model need not be complete, and can be interpreted however the student wishes, so long as there is justification for each decision.

An example of a model of the human brain could be:

- A series of computers, each programmable to different functions of the human brain, joined together with a series of wires acting as the neuronal connections between each component.
- Limitations of a system such as this could be size, as the number of computers required would be too numerous to describe, and not at all proportional to the human brain. We are also limited by our human understanding, both of technology and computing, and of the brain itself. Brains are adaptable and have 'plasticity' – a computerized model would not be able to show this feature, since they will have to be programmed in advance of their function. In this way, any model of a human brain would be reductionist.

Any other model of a brain is credible and equally successful so long as it is logical and key components are described and justified in detail.

2. *"Impatience is the driving force behind development." To what extent is patience no longer a virtue in modern society?*

Introduction:

- Consider a definition of patience – 'the capacity to accept or tolerate delay, problems, or suffering without becoming annoyed or anxious' and impatience as the opposite to this.
- You may wish to explore the phrase 'patience is a virtue' – where a virtue is a character trait which is considered good or desirable.
- Make a reference to the original quote – which suggests that it may be impatience that is valued with the way that society is rapidly changing and developing. Emphasis on 'modern' society – is the phrase outdated?
- Introduce ideas on both sides of the argument.

Possible arguments for:

- Arguable that to be successful, you must be impatient for change and success – if you aren't, you aren't passionate enough and you won't succeed. Good leaders are driven and motivated for change, which often comes from an impatience.
- The speed that technology (for example) is advancing and changing means that you have to be impatient for the next thing otherwise you will always be one step behind.
- Learning is also helped by impatience – the more you want to know NOW, the harder you will work to understand it. Impatience at your current capabilities and understanding will lead to a drive a passion to learn and comprehend more.
- Impatience is perpetual – the quicker we get what we want, the quicker we want it next time. Therefore impatience is required to stay driven and committed.

Possible arguments against:

- Impatience leads to stress, which can have several impacts – on health, on relationships, on work. Therefore the consequences of impatience can be detrimental and it is undesirable to be impatient.
- Patience leads to calmness and relaxation, which are important for the ability to rest our bodies and taking time to enjoy the world. If we forget how to be patient, we will stop having time for ourselves, and, even worse, time for other people.
- We have no need to be impatient in such a fast paced society – things are changing so quickly, we can access what we want so quickly with technology that impatience is unnecessary and patience should be practiced before we forget how to employ it.
- Patience allows us to value the world – it gives us time to think, and stops us taking everything for granted.
- Patience is a character trait which is looked for in friendships – everyone wants a friend who they feel will listen and will spend time with them, not someone who is constantly in a rush for the next thing to happen.

Conclusion:

- Summarize the main points made on each side of the argument in the essay.
- Link back to the original quote – mention development and drive.
- You may wish to come to a decision either way, or it is equally fine to sit somewhere in the middle, so long as this is fairly justified.

3. *Voltaire said, "The perfect is the enemy of the good." Discuss this claim.*

Introduction:

- Consider a definition of perfect - having all the required or desirable elements, qualities, or characteristics; as good as it is possible to be, versus good - to be desired or approved of, of a high standard.
- You may wish to discuss the term 'enemy' – by this it something that is directly opposed to something else.
- Best to consider this in the context of striving for each quality – is either ever achievable?
- Introduce the points you will make on both sides of the argument.

Possible arguments for:

- Perfection is unachievable in a world that has so much wrong and badness in it – therefore striving to be perfect is impossible and will only lead to disappointment.
- Goodness cannot be appreciated fully if the ultimate goal is perfection, because everything below that goal will not be good enough. Therefore a state of perfection eliminates the possibility of goodness because nothing, other than perfection, is 'good enough'.
- Striving for perfection causes inevitable failure, and failure is not desirable, and (since good is defined as something to be desired), therefore it is not good.
- Example of clinical perfectionists – unhappy, constantly working, can lead to or be connected with mental health issues.

Possible arguments against:

- Perfection and goodness go hand in hand, while striving for goodness, you are getting closer and closer to perfection. Therefore perfection aids goodness.
- Things can only look good in relation to the superlative – so how could we judge goodness without perfection?
- Consider a sliding scale between perfect and imperfect – good is closer to perfect, and therefore cannot be considered its enemy.
- If perfection is unachievable, but goodness is, you can find goodness in the pursuit of perfection, so long as you allow for something to be 'good enough'.
- Perfection is 'ultimate goodness', but goodness is only 'not good enough' if we decide it to be so – it is all a matter of perspective and goal setting.

Conclusion:

- Summarize the main points on both sides of the argument.
- Link back to the original quote – you may wish to bring up the idea of friends and enemies again.
- You may choose to ultimately agree or disagree with the claim, or sit in the middle with justification on either side – so long as the argument is logical following from the rest of the paragraph, either is equally credible.

4. Is world peace achievable?

Introduction:

- Consider a definition of peace- freedom from disturbance, tranquillity, the absence of conflict. Or relate this directly to world peace - freedom, cooperation and happiness among and within all nations and people.
- It may be relevant to mention the number of wars and the amount of conflict going on around us at the moment – both in wars and more silently, in areas of corruption and poverty.
- You could introduce the ways in which it could be reached – either voluntarily or by virtue of a system of governance that prevents warfare. Is the latter true peace?
- Introduce the ideas on both sides of the argument you wish to discuss.

Possible arguments for:

- Achievable is a goal that is possible – and it is certainly possible for all nations to cooperate, given the correct guidance and when there is will on all sides
- The first step is believing it is possible, and we do not know the future. An 18th century European would never believe that German, French, Austrians etc could ever live peacefully together, but this has happened. When there is constant change, we cannot decide that anything is impossible. When there is a will, there is a way.
- So long as there is a dream and the concept is still alive, there is always the possibility that all nations can live happily together. Cooperation can take place, people can unite.
- The world is made up of people, each with their own thoughts and their own beliefs. We are all human, we all make mistakes, but there is also good in everybody and when striving for a common goal, that can be reached.
- The importance should be placed on a strategy and an ultimate goal – a system working together, involving individual people joining as one, such as in government, can always work for peace. We have the quality within us so long as we choose to use it.

Possible arguments against:

- It is easy to get carried away with dreaming – just because something is possible doesn't always mean it is achievable? It is not realistic to think that we can solve such a massive issue.
- There are too many variables involved – religion, economics, race, etc and too many opinions. How could we achieve everyone seeing eye to eye when everyone has different opinions?
- Look at history – as one war ends, another one begins. We learn from the past, and the past has shown that humans cannot live harmoniously with each other.
- As long as there is badness in the world – sin and corruption- in the form of greed, impatience, anger, selfishness – there will always be conflict. And world peace would be achieved if there were no conflict, so it is not achievable.

Conclusion:

- Summarize points made on both sides of the argument concisely and draw ends together.
- Link back to the original point of world peace as being an ideal, and ultimatum – achievable meaning something can be reached as a goal. Ensure all arguments are relevant to this point.
- It is credible to either come to a conclusion for or against world peace, or to come to a compromise – such as world peace being possible as a goal, but with conditions or some consideration for conflict or argument.

END OF PAPER

Mock Paper E: Section 1

Question 1: C
The simplest solution is to calculate the total area at the start as 20 x 20 = 400cm^2. Then recognise that with every fold the area will be reduced by half therefore the area will decrease as follows: 400, 200, 100, 50, 25, 12.5 – requiring a total of 5 folds.

Question 2: D
This is the only correct as it is the only statement that doesn't categorically state a fact that was discussed in conditional tense in the paragraph.

Question 3: B
Off the 50% carrying the parasite 20% are symptomatic. Therefore 0.5 x 0.2 = 10% of the total population are infected and symptomatic. Of which 0.1 x 0.9 = 9% are male.

Question 4: D
The most important part of the question to note is the figure of 30% reduction during sale time. Although A and B are possible the question asks specifically with regard to cost. Therefore, it is only worth waiting for the sale period if the sterling to euro exchange rate does not depreciate more than the magnitude of the sale. As such solution D is the only correct answer as it describes anticipating a loss in sterling value less than 30% against the euro.

Question 5: D
Begin by calculating the number of childminders that can be hired for a 24-hour period as 24 x 8.5 = 204. Therefore, a total of 4 childminders can be hired continually for 24 hours with £184 left over – as the question states the hire has to be for a whole 24-hour period and therefore the remainder £184 cannot be used. As such D is the correct answer of 4 x 4 = 16.

Question 6: B
The simplest way to approach this question is to recognise that there is a difference of £1.50 between peak and off-peak prices for all individuals except students. The total savings can therefore be calculated as (3 + 5 + 1) x 1.5 = 9 x 1.5 = 13.5.

Question 7: A
Karen is a musician, so she must play an instrument, but we do not know how many instruments she plays. Although all oboe players are musicians, it does not mean all musician play the oboe. Similarly, oboes and pianos are instruments, but they are not the only instruments. So, statements b and c are incorrect. Karen is a musician but that merely means that she plays an instrument, we do not know if it is the oboe. So, statement d is incorrect.

Question 8: D
Answers A and B are simply incorrect as the measurement taken is a percentage increase (/decrease) which will normalise baseline diameters therefore allowing for comparison over multiple time points. You should be aware from your studies that ultrasound is an invaluable technique in distinguishing between adjacent tissue types. Any methodology is repeatable if it is correctly chronicled and followed therefore leaving the correct answer of D.

Question 9: C
If both the flight and travel from the airport are delayed this will be the longest the journey could possible take – producing a total journey time of 20 + 15 + 150 + 20 + 25 = 230 minutes or 3 hours 50 minutes. Therefore given all possible eventualities, to arrive at 5pm, boarding should begin at 13.10pm. Answer D is incorrect as a delayed plan would add 20 minutes to the journey whilst the transport to the meeting at the other end takes a minimum of 15 minutes – even if Megan could teleport instantaneously from the airport to the meeting she would be 5 minutes later than if there wasn't a plane delay.

Question 10: E
This is almost a trick question and simply an application of exponential decay. Recall that an exponential decay is asymptotic to 0 as no matter how small the volume within the cask becomes, only half of it is ever removed. It could be argued that this process cannot continue once a single molecule of whiskey is left – and when splitting that single molecule in half it is no longer whiskey. However, the question does not ask "how long till all the whiskey is gone" but rather "how many minutes will it take for the entire cask to be emptied" and therefore the process can continue infinitely – even if the only thing left in the cask is a collection of quarks … or half that.

Question 11: D
A sky view of the arrangement leads to:

| C | A | B | | | A | C | B |
| D | | E | *or* | | D | | E |

In both, D is to the left of E, thus is the only correct answer.

Question 12: E
The question can be expressed as $(40 \times 30) - x(50 \times 30) = 200 = 1{,}200 - x1{,}500$. Therefore $x = 2/3$.

Question 13: D
The information given is very much a red herring in this question. This can be solved using your own application of sequence theory. If a sequence is 6 numbers long, the 6 numbers can be re-arranged into a total of 6! possible sequences $= 1 \times 2 \times 3 \times 4 \times 5 \times 6 = 720$

Question 14: E
Although this is an extremely abstract question, all the information needed to answer it is provided. The key rule to have isolated from the information is that with each progressive sequence, a bell can only move one position at most. Therefore, looking through each in turn – for example we can exclude A as the 3 goes from position 4 to position 2. This leaves the only correct answer of B. Recall that the information states bells can only move a maximum of one place – they can of course move 0 places.

Question 15: A
As the largest digit on the number pad is 9, even if 9 was pressed for an infinitely long time the entered code would still average out at no larger than 9. Therefore, it would be impossible to achieve a reference number larger than 9. Indeed, this is an extremely insecure safe but not for the reason described in B (for if the same incorrect number was pressed indefinitely it would never average out as the correct one) but rather because the safe could in theory be opened with a single digit.

Question 16: D
A is incorrect as it ignores the section of the text that states the evolution of resistant strains is driven by the presence of antibiotics themselves. The text states that the rate of bacterial reproduction is a large contributing factor and therefore not wholly responsible – hence B is incorrect. Since this is just one example (and only the information in the text should be considered for these questions) for C to make such a general statement is complete unjustified.

Question 17: B
The fastest way to solve this question is to calculate the quantity of cheese per portion as $200/10 = 20$. Which for 350 people would require $350 \times 20 = 7000g$ or 7kg.

Question 18: C
Calculate the calorific content of 12 portions as $12 \times 300 = 3{,}600$kcal. As this represents 120%, evaluate what the initial amount would be as $(3{,}600/120) \times 100 = 3{,}000$kcal.

Question 19: B
Begin by calculating the initial weight of all the ingredients in the Bolognese sauce which comes to a total of 3.05kg. Therefore when cooking for 10 people 3.05 x 4 = 12.2kg of pasta should be used. Which in turn means for 30 people 3 x 12.2 = 36.6kg should be used.

Question 20: D
Calculate the new weight of ingredients in the Bolognese sauce excluding garlic and pancetta which produces a total of 2.8kg. Note that onions represent 0.3kg per 10 people and as such the ratio can be represented as 0.3/2.8 or alternatively dividing top and bottom by 0.3 → 1/9.3

Question 21: E
Begin with calculating total preparation time as 25 x 4 = 100 mins. The fact that Simon can only cook 8 portions at a time is somewhat a red herring as it doesn't impact the calculation. Total cooking time can be calculated as a further 25 x 8 = 200 mins. Producing a total time of 300mins or 5 hours.

Question 22: D
Due to the quantities colour t-shirts are priced at £5 and black and white at £2.50. Therefore, the order will incur a total cost of 50 + (50 x 5) + (200 x 2.5) = £800.

Question 23: E
Answers A and B directly conflict with information presented in the text whilst C and D may well be true but there is insufficient information in the text to address these points. This is an important reminder that although you may well have been a scout and be able to comment on options C and D, you may only consider these arguments in terms of the text provided.

Question 24: B
There are two directions: clockwise and anticlockwise and rats will only collide if they pick opposing directions.
Rat A - clockwise, Rat B - clockwise, Rat C - clockwise = 0.5 x 0.5 x 0.5 = 0.125.
Rat A - anticlockwise, Rat B - anticlockwise, Rat C - anticlockwise = 0.5 x 0.5 x 0.5 = 0.125.
0.125 + 0.125 = 0.25. So, the probability they do not collide is 0.25.

Question 25: C
The article seen here is a particularly good effort at a discursive text as it is completely impartial. Note that the article simply states the facts from either side in equal measure. Nowhere does the author present their opinion on the matter nor do they insinuate their beliefs in anyway.

Question 26: C
The journey time is rounded to the nearest hour (13). Therefore, the longest it could possibly be is 13 hours 29 minutes or it would be round up to 8 hours. Therefore, the latest the ferry will arrive, assuming the travel time estimate is accurate, is given as 20.29.

Question 27: B
The correct answer is 28. Assume, although very unlikely, that you roll a 2 every single go – you will never need to take a step back, only your two forward. Therefore, rolling a lower number in this case is beneficial.

Question 28: B
The simplest way to calculate this is to find the lowest common multiple of the given laptops which is 40 x 60 x 70 = 168,000 seconds = 2,800 minutes = 46.67 hours – although they will not all be on the same lap number at this time.

MOCK PAPER E SECTION 1 — ANSWERS

Question 29: C
A large amount of subtly different data is described here. Of note is the first experiment which describes how nerve conduction is faster in right handed men than it is in left handed men. This result is not transferable to women until it is proven! The experiment currently being conducted only considers dominant hand in men vs. women. That could be either hand or not necessarily the females' right hand. For example, all of the females in this experiment could have been left handed and there is no information in the text to say otherwise, therefore we cannot tell.

Question 30: C
Recognise that two square based pyramids will comprise 8 triangles of base width 5 and height 8; plus, two 5 x 5 squares. Thus, giving at total area of 8(8/2 x 5) + 2(5 x 5) = 210 cm^2

Question 31: B
In order to approach this question first realise that in the first well 1ml of solvent is being combined with 9ml of distilled water producing 1eq of solute in 10ml – hence the first well produces a dilution by a factor of 10. With each progressive dilution the concentration is reduced by a further factor of 10 – hence by well 10 the concentration is at $x/10^{10}$

Question 32: E
The compartments of the human body are occupied by numerous fluids, as the student is only interested in measuring the volume of blood, it is essential he chooses a solute that will only dissolve in blood. So as his known quantity of solute remains no, it must be neither removed nor added during its time in the body. Hence all of the written assumptions must be made and many more.

Question 33: D
The fastest way to approach this question is by calculating the total price per head for the cheapest option as 10 + (8/20) + (10/60) = 10.567. To make the maths simpler this can be rounded safely to £11 a head at this stage. 2,300/11 is approximately equal to 209 which when rounded to the nearest 10 is 210 people.

Question 34: A
Whilst this passage is attempting to weigh up two sides of an argument, it has a clear one-sided approach focussing heavily on the excitement of dangerous sports. It even states that hunting is recognised as exciting by some. Since the previous sentence discussed the link between archery and hunting, the statement is a fair extrapolation to make.

Question 35: A
If society disagree that vaccinations should be compulsory, then they will not fund them. So, statement A is correct. It attacks the conclusion. Statement b - society does not necessarily mean local so this does not address the argument. Statement c strengthens, not weakens, the argument for vaccinations. Statement d – the wants of healthcare workers do not affect whether vaccinations are necessary.

Question 36: C
Start by calculating the area of wall that may be painted per tin of paint as 10 x 5 = 50m^2. Therefore, to paint the whole area 1050/50 = 21 tins of paint are required per coat. As such to complete 3 coats it will cost Josh 3 x 21 x 4.99 = 314.37.

Question 37: C
A is a correct assumption as procession is a function of rotational motion. B is a necessary assumption or rather inference of the first sentence. The second sentence only says that an asterism can be used, not that it is the only possible method. Nothing is mentioned of navigating the Southern Hemisphere and therefore C is not a valid assumption.

MOCK PAPER E SECTION 1 — ANSWERS

Question 38: D
Recognise that "bank hours" refers only to hours that the bank is open – which Mon to Fri is 8 hours whereas it is only 6 hours on a Saturday. Although John needs the money by 8pm the bank closes at 5 and that 3 hours difference cannot be used. Hence working backwards John will need 8 hours on the Tuesday, 8 hours on the Monday, Sunday is closed, 6 hours on the Saturday, 8 hours on Friday and 8 hours on Thursday and 4 hours on the Wednesday. With a closing time of 5pm, the latest John can cash the cheque on Wednesday is 1pm.

Question 39: D
First thing to recognise here of course is that individual diamonds can be combined to form larger diamonds with the 5 x 5 diamond the biggest of them all. To avoid counting them all and risking losing count, instead deduced the number of triangles per corner and per side; then multiply up by 4.

Question 40: B
Let my current age = m and my brother's current age = g. The first section of this question can therefore be expressed as $m + 4 = 1/3(g + 1)$ whereas the second half can be represented as $2(m + 20) = g + 20$. Therefore, this problem can be solved as simultaneous equations. Rearranged the second equation reads $m = 1/2 g - 10$; when substituted into the first equation we form $1/2 g - 10 + 4 = 1/3(g + 1)$. Expand and simplify to $1/2 g - 6 = 1/3 g + 1/3$ → $1/6 g = 6\frac{1}{3}$ which therefore means my brother's current age $= 6\frac{1}{3} / (1/6) = 114/3 = 38$. Which means that my current age $= 1/2(38) - 10 = 9$.

Question 41: D
A is categorically wrong as the first two paragraphs discuss how aneurysms produce inflammation which in turn blunts endothelial NO action. B is incorrect as it states aneurysms directly promote CVD, this is not a direct process. It is the blunted NO which directly produces the CVD. C can be ignored as nowhere are aneurysms categorised like this. E is incorrect as the text states that aneurysms reduce NO which will reduce vasodilatation, thus increasing basal vasoconstriction and thus reducing blood flow. Leaving the correct answer of D which is of course true as observations are not transferable between species until tested scientifically.

Question 42: C
Any statement which refers to national or global figures is instantly incorrect as the text does not mention any statistical analysis has taken place. In order to produce national statistics from a small sample size such as this requires statistical analysis. Whilst E could possibly be true it cannot be stated as there are so many possibilities – perhaps the time of the survey was during rush hour in which case the majority of the traffic would have been travelling in the same direction anyway to reach an industrialised area.

Question 43: B
The runners aren't apart at a constant distance; they get further apart as they run. Xavier and Yolanda are less than 20m apart at the time William finishes. Each runner beats the next runner by the same distance, so they must have the same difference between speeds. When William finishes at 100m and Xavier is at 80m. When Xavier crosses the finish line then Yolanda is at 80m. We need to know where Yolanda is when Xavier is at 80m. William's speed = distance/time = 100/T. Xavier's speed = 80/T. So, Xavier has 80% of William's speed. This makes Yolanda's speed 80% of Xavier's and 64% (80% x 80% = 64%) of William's. So, when William is at 64m when William finishes. 100m - 64m = 36m, thus William beats Yolanda by 36m.

Question 44: A
This question can be solved quickly if you first realise that there is no need to calculate both volumes and subtract the larger from the smaller, instead only convert the television dimensions into metres and then calculate 60% of that.

Question 45: E
From the information provided all the flaws listed are valid since David's main point is that he has chosen the cheapest. A could be true as there is an additional cost of £3 for staying at Whitmore, therefore if the vehicle they are using achieves sufficient miles per gallon then travelling the extra few miles could cost less than £3 in terms of petrol. B again is possible which would argue against it being cheap, as would D. And if C is true then David's argument is flawed altogether.

Question 46: D
C is irrelevant as nowhere does the passage mention standards of modern medical practice. A may be incorrect as nowhere does the article explicitly say that animal testing is the only accepted method of drug approval. B categorically conflicts with the first sentence of the second paragraph.

Question 47: A
Begin by converting all the quantities into terms of items as that is the terminology used on the graph axis. Therefore 12 rugby balls = 6 items and 120 tennis balls = 24 items. Reading from the graph reveals their respective prices as £9 and £5. Therefore, the total cost of products in the order is (6 x 9) + (24 x 5) = 174. Since this is significantly more than £100 the delivery charge is waived.

Question 48: D
Calculate the cost of 10 of everything as (2 x 5) x (10 x 7) x (5 x 9) = £125. Recall that delivery charge is waived at £100 and this therefore a trick question and no delivery charge is applied anyway.

Question 49: E
Tennis balls are sold in the largest pack and so they must be considered. Begin by dividing 1000/5 using the value from the first column = 200. As this is above the range 0 -99 look up the item value in the 100 -499 range where a £1 discount is applied per item. Therefore, in actually fact 1000/4 = 250 items can be purchased which equates to a total of 250 x 5 = 1250 balls.

Question 50: B
Recognise that 120% profit is equivalent to 220% of the original price. In which case the initial purchase price = (1,320/220) x 100 = £600.

END OF SECTION

Mock Paper E: Section 2

1. *"Nature over nurture".* **To what extent do you agree with this statement?**

Introduction:
- Define briefly what 'nature' and 'nurture' entail – for example, nature can refer to natural attributes such as hereditary traits (e.g. IQ, height, eye colour) whilst nurture can refer to one's environment and circumstances
- State your standpoint as to why you believe nature trumps nurture or vice versa
- Example standpoint: I believe that nurture trumps nature as successful people tend to be brought up in favourable conditions and are surrounded by encouraging and inspirational people. These conditions all constitute 'nurture' and play a larger role in determining one's success as opposed to what natural traits the person was born with (e.g. looks, intelligence)
- Example alternative standpoint: 'Nature over nurture' is largely true due to the fact that even though one's circumstances play a role in determining one's success, ultimately one's genetic traits play a disproportionately larger role in determining a person's pathway in life.

Possible arguments for nurture over nature:
- Successful people tend to be brought up in favourable environments and are surrounded by motivational people
- For example, many successful musicians tend to be brought up in a family tend encourages the study of music, or had parents who were successful musicians themselves
- It is well-known that more well-to-do families tend to produce more well-adjusted children who have a statistically higher chance of earning more in their working life
- Peer influence plays a huge role in the development of a child – e.g. if a child hangs out with children who constantly skip school and engage in gang violence, the child is most likely going to be negatively influenced
- Talent is not enough without hard work – successful artists, musicians and scientists had to work very hard to get to where they are even if they had some talent to begin with
- Studies have shown that a person's characteristics are determined by a combination of genetic factors as well as environmental factors – the study of epigenetics shows that environment plays a huge part in determining what traits a person develops

Possible arguments for nature over nurture:
- Successful people like Einstein, Bill Gates, Tim Cook etc. were all naturally brilliant and clearly had the intellectual capacity to achieve great things, and intellectual capacity is very much a hereditary trait
- Successful actors and models are largely successful due to their good looks, again another hereditary trait
- Studies have shown that good-looking people tend to progress faster at work as they are naturally more confident and are better at networking and establishing good client relationships
- Having more natural intelligence also means that these people can succeed in life much easier with less hard work
- If you are born with a serious defect, or if you were born with deformities, you are naturally going to be much more hindered in life compared to luckier individuals who were born with no conditions
- Even though genetic deficiencies and conditions can be overcome, it does make one's life harder and the starting point in life is drawn largely based on one's luck in the genetic lottery
- Studies have shown that psychopathic criminals tend to have deficiencies in certain chemicals that result in a different expression of mood in their brains, which shows that serious criminals are not produced based solely on their circumstances

Conclusion:
- Include a summary of all points given and refer back to your original definition and standpoint
- You do not have to come up with an extreme conclusion (e.g. nature always trumps nurture, or vice versa), a sensible conclusion that is measured and discusses the possible exceptions is perfectly acceptable

1. Do you think positive discrimination should be implemented in our university admission process?

Introduction:

- Define positive discrimination – the act of lowering entry requirements for certain minority groups in order to overcome inequality and disproportionate representation and improve social mobility
- You may give refer to particular examples such as the US implementing this in their university admission process
- State your standpoint – whether you agree that this should be implemented in order to improve the representation of minority groups in our universities, or whether this goes against meritocracy and produces more harm than good

Possible arguments for positive discrimination:

- Certain minority groups such as black students come from less well-to-do families and do not have access to the same resources in order to do well for their examinations
- Minority students are less encouraged to apply for top universities because they perceive these universities as elite and lacking representation of minority groups
- Minority students do not receive adequate support and consideration for differences in their background and community
- Diversity in university education is important in order to foster a more vibrant community
- It is important to allow more disadvantaged minority students to enter top universities if they have the intellectual capability as this will improve social mobility by improving their chances of securing top-paying jobs upon graduation
- Many minority students are given a lower starting point in life due to inherent discrimination and stereotypes in society, as well as their less well-to-do backgrounds, hence positive discrimination is needed in order to level the playing field
- Universities benefit from having a larger representation of minority students amongst their student population as they bring in different ideas, perspectives and thoughts, which stimulate intellectual academic discussion and studies
- Studies have shown that diversity leads to greater productivity and fosters creative thinking and innovative solutions
- Admissions should not be based solely on raw grades as this does not take into account the fact that many minority students might not have the same resources or teaching in order to score as well as they are capable of in the examinations

Possible arguments against positive discrimination:

- We should not lower entry requirements for a particular group just because they happen to be a minority group as this goes against the idea of admissions based solely on merit
- People might start to have the perception that minority students only managed to enter some top universities due to their race – a situation that is happening in the US
- Majority groups may feel that it is unfair that they failed to get into a university compared to a minority student even though they have the same grades just because of their race
- Lowering entry standards may result in some students entering universities they are not suitable for academically
- Positive discrimination is a slippery slope and might lead to identity politics – whereby students become increasingly polarised and segregated based on their identity
- Many of our top universities pride themselves in being meritocratic and only accepting the best of the best – arguably lowering entry requirements just for a particular group of students risk harming our high standards
- Instead of implementing positive discrimination, we should instead focus more on tackling the problem of social inequality and social mobility by ensuring that minority students receive the education they deserve at the lower levels

Conclusion:

- Include a summary of all points made, and give a balanced overview of both sides of the argument.
- Reiterate what your overall conclusion is and why you have arrived at that conclusion
- A sensible and measured conclusion is desirable – if you have a strong viewpoint, be prepared to defend it

3 **'Developed countries have a responsibility to help developing countries grow'.
Do you agree with this statement?**

Introduction:

- Define the difference between developed and developing countries
- Give some examples – e.g. developed modern countries such as the US and Japan vs. developing countries that require infrastructural support such as certain Southeast Asian countries and African countries
- State your standpoint – why should developed countries help developing countries grow, or why not?

Possible arguments for:

- Many developing countries managed to develop at a fast pace by exploiting the resources from developing countries
- E.g. Colonialism was known to have helped many countries such as the UK, Spain and Portugal develop quickly at the expensive of the colonised countries
- It is also in the interest of developed countries to help developing countries grow due to the interconnectedness of economies these days
- With greater economic growth, developing countries will experience less instability issues and this will result in greater world peace, which is of benefit to developed countries
- Developed countries have greater resources and have a responsibility of sharing these resources with developing countries in order to help them develop and lift their population out of poverty
- Economic growth is not a zero-sum game and it does not mean that if developed countries help out developing countries, they reduce the amount of resources available for their own people – on the contrary, helping developing countries grow has tremendous knock-on effect in the long run by improving trade relations and diplomatic relations between the two countries
- Developing countries tend to have an abundance of natural resources and developed countries stand to benefit from such resources if they are willing to help developing countries with their infrastructural growth
- In terms of a humanitarian standpoint, many developing countries suffer from acute problems such as lack of clean water, food and resources, and developed countries have a humanitarian duty to do their best in helping these countries and ensuring that people have access to clean water and sanitary conditions
- Having developed countries help developing countries also reduces the risk of inter-country conflict and war as they become more dependent on each other

Possible arguments against:

- Developed countries have their own problems to take care of (e.g. slowing economic growth, strained infrastructure, congestion problems etc.)
- Developed countries may not know how to help developing countries due to the intricacies of their problems
- Sending aid to developing countries does not always work – e.g. sometimes developing countries become overly-dependent on the aid
- Developing countries need to be more independent in developing on their own pace and manner in order for them to be sustainable in the long run
- There have been situations where developed countries have made developing countries worse off by attempting to help them – for example causing more pollution and damage, or stagnating the developing countries' growth by making them heavily-reliant on aid
- Resources are scarce and developed countries can only do so much to help developing countries that are miles away without compromising on the attention they need to give to their own citizens
- Developing countries need to have more effective governance in order to lift their own people out of poverty – this is something that developed countries have little control over due to sovereignty issues
- If the government of a developing country is heavily-corrupted, or if the country is highly sealed off from the world and insular, there is not much developed countries can do to help the country without risking heightened conflict

Conclusion:

- Include a summary of all points made, and give a balanced overview of both sides of the argument.
- Reiterate your overall conclusion and explain how you arrived at your conclusion
- Provide a sensible and measured overall conclusion based on the arguments and counter-arguments you have raised

4 Should our education system place a greater emphasis on the sciences as opposed to the arts?

Introduction:

- Define both the sciences and arts and explain how an education in both might differ
- Contrast the different subjects considered – e.g. learning about physics, biology and chemistry in the sciences vs. learning about art, history and literature in the arts
- State your standpoint – should more emphasis be given to the sciences or should the arts not be neglected?

Some points as to why science should be given a greater emphasis:

- Science is experimental – involving accurate and detailed study.
- Science could be argued to be physical and measurable- using known concepts in the real world that can be objectively identified.
- Science is adaptable and ever changing – according to research and current thought.
- Science could be considered to be held within the thoughts and brains of society – if there weren't people exploring it or interested in it, would it really exist?
- Science involves logical and critical thought – in that sense it could be considered to be merely a thought process – but you may wish to explore the idea that it couldn't exist without physical things to measure in themselves.
- Valuable for the evolution of society – as we learn more about the world and about ourselves, this helps change to occur which is vital.
- Provides concrete fact and stability.
- Gives a purpose to life, which brings fulfillment and contentedness.
- Brings explanation for some things that can't be explained through simple observations, which allows a firmer foundation and an answer to bigger, crucial questions.
- Helps us to gain more of an understanding of the world we are in – bringing a stronger grasp on reality and greater knowledge and insight.
- Medical reasons – drugs, medicines, health.
- Safety – e.g. monitoring volcanoes, predicting earthquakes etc.

Some points as to why arts should not be neglected:
- Learning the arts teaches us how to express ourselves
- Learning the arts foster creative thinking
- Arts place an emphasis on critical thinking
- Learning about history and philosophy can be useful in shaping political and critical thought
- Effective writing and communication is essential and can be picked up from the arts
- Arts comprise a wide variety of subjects – ranging from literature to history to music – all these subjects inculcate different skills and many students have differing levels of aptitude and interest that can be catered by the different subjects
- Many successful politicians and government officials studied arts subjects – clearly being good at an arts subject can be indicative of certain useful attributes such as being an effective communicator, being a critical thinker and being able to see the wider picture
- While the sciences are important, learning the arts help to maintain and develop our culture and helps students develop an appreciation of the finer things in life
- Languages are also increasingly important these days within our globalised world – whereby multilingual individuals are highly favoured in the workforce and a premium is available
- In this day and age, technical abilities are not the only thing employers look out for – communication and effective writing skills are also highly important and these are adequately trained by arts subjects

Conclusion:

- Summarize the key ideas explored previously – the ideas associated with science and arts and their relative importance
- Ensure to link back to the original statement – should more emphasis be placed on the sciences in our education system?
- Provide a good summary of the arguments and counter-arguments you have raised and why you have arrived at your overall conclusion

Mock Paper F: Section 1

Question 1: B
Lucy must live between Vicky and Shannon. Lucy is Vicky's neighbour, so Shannon cannot have a red door. Vicky lives next to someone with a red door, so Lucy must have the red door. This leaves Shannon with the blue door and Lucy with the white. The green door is across the road and so does not belong to any of them.

Question 2: E
First calculate an average complete one-way journey time as 40 + 5 + 5 = 50 minutes. Deducting his breaks, he works a total of 7 hours 20 or 440 minutes. Since the first train is already loaded his first run will only take 45 minutes leaving 395 minutes to complete his working day. 395/50 = 7 remainder 45. Note that 45 minutes is not enough to fully unload the train, but it is enough to load the train and drive the distance. Therefore, the driver will complete a total of 9 journeys equalling a distance of 198 miles.

Question 3: E
A is not actually a valid assumption as we do not know what proposal conservationists might be bringing to the local councils, they have only expressed their concern. They may well be bringing a proposal to ask for funding to rehome all the species in the affected environment. B is essential to the final paragraph whilst C must be assumed otherwise the councils would not be presenting these proposals at all.

Question 4: D
As there is not really information in the question to calculate the answer quickly. Instead consider each answer in term and calculate the differences to find the correct price difference in the question:
   A) (3x 1) + (2 x 1.25) – (15 x 0.3) + (10 x 0.5) etc...

Question 5: E
Based on the information in the question options A - D are simply wrong. A is incorrect as antibiotic E has not affected growth at all. B is incorrect as the other antibiotics have significantly affected growth. E was the least effective antibiotic. C was not the most ineffective as it did disrupt growth slightly whereas E had no effect at all. D will now be taken and the experiment repeated with D at numerous concentrations to find the optimum dose.

Question 6: A
1L = 1000 cubic centimetres and therefore the total volume of air Laura needs to produce is 25 x 0.3 = 7.5L. With a total of 25 balloons she will take 25 x 0.5 = 12.5 seconds breathing in and a further 7.5/4.5 = $1\frac{2}{3}$ minutes inflating the balloons. This yields a total time of 1 minute 40 + 12.5 seconds = 1 minute 52.5 seconds or 112.5 seconds.

Question 7: B
Quickly represent the question schematically as (A = B) ≠ (C = D = E). We can now observe that A in fact supports George's argument, C also supports George's argument and D may well be true, but it would have no effect in disrupting the argument, simply only imply D and E are both also equal to 0. However, as E is equal to C it should therefore not equal B.

Question 8: B
This question is much less complicated than it sounds. Begin by just considering a single hour. Throughout the hour of 1 the hour hand will be pointing at 1. Only during the 5th minute of that hour will the minute hand point to the 1 whereas every 5th second of the minute the second-hand points to the 1. All these events will only coincide once. As there are 24 hours in a day 00:00 through to 23:59 this event will happen 24 times.

MOCK PAPER F SECTION 1 — ANSWERS

Question 9: B
A) Potentially correct, but extreme sports also carry higher risks of injury.
B) True.
C) True, but irrelevant for the question.
D) Potentially correct, but irrelevant to the question.

Question 10: D
A) False – we are not told about the healthcare directly but are told that injury and disease posed a threat.
B) False – the terrain was difficult, and mapping was poor.
C) False – outlaws were a significant threat.
D) True – as the text states, there was a marked lack of bridges.

Question 11: C
Whilst B may be true it is not a reason for dependence, only a supporting factor. Dependence implies that we have no choice but to use electricity. Hence A is wrong as gas is readily available; hence D is wrong for the same reason. This leaves the correct answer C which is the only statement which truly describes our absolute necessity for electricity – since electrical appliance by definition only function with electricity.

Question 12: B
First note that 27 guests plus Elin herself means that 28 people will be eating the 3 courses which will require a total of 28 x 3 = 84 glasses of wine. This is a total volume of 84 x 175 = 14.7L = 21 bottles. As wine is only sold in cases of 6, Elin will have to buy 24 bottles so as not to run out. Recall the buy one get one free offer so she only pays for 2 cases.

Question 13: E
Recognise that when rounded to the nearest 10 the shortest an episode could last is 35 minutes. Hence a total of 7 x 12 = 84 episodes would take a total of 84 x 35 = 2,940 minutes = 49 hours.

Question 14: E
Points A and B are the best exemplified through this passage. Often great discoveries come from accidental observations and then exact processes are refined through many experiments in a trial and error fashion until the correct methodology is achieved. The passage demonstrates how as our understanding of the world around us advance so too does our ability to provide healthcare. D can be observed in the passage as the 50/50 split.

Question 15: C
Whilst A and D are true they do not force the stranger to give him the sapphire – remember Jack can be given any stone for a truthful statement. B and C are both lies and will earn Jack nothing. Instead if Jack states E then the stranger has no choice to hand over the sapphire else it would be a false statement.

Question 16: A
Despite the enormous interest rate in Simon's current account it is only awarded twice, whereas in the saver account it is awarded 4 times. Hence earnings from the saver account = 100 x 1.5^4 = £506 whereas earnings from the current account would have stood at £361.

Question 17: D
The largest possible key can be obtained were the first two numbers are at a maximum because they are multiplied together → 9 x 9 = 81. Subtract the smallest number to yield 81 – 1 = 80 and again divide by the smallest number which is 1 hence 80 is the largest possible key.

Question 18: D
A) Incorrect. The text clearly states that the exercise routine is resistance training based.
B) False. Both groups contain equal numbers of men and women per the text.
C) False. Both groups are age matched in the range of 20 to 25 years.
D) Correct. As the only difference between the two shakes is the protein content.

MOCK PAPER F SECTION 1 — ANSWERS

Question 19: D
There are three different options for staying at the hotel. They could either pay for three single rooms for £180, one single and one double room for £165, or one four-person room for £215.

Subtracting the cleaning cost for one night would leave:
£180-(3x£12) = £144
£165-(2x£12) = £141
£215-£12 = £203

The cheapest option is one single and one double room and they want to stay three nights, giving £141x3 = £423.

Question 20: D
Glass one starts with 16ml squash and 80ml water. Glass two starts with 72ml squash and 24ml water. 48ml is half of 96ml so 8ml squash and 40ml water is transferred to glass two. Glass two now contains (8+72 = 80ml squash) and (24+40 = 64ml water). Glass two now has a total of 144ml and half of this is transferred to glass one. Glass one now has (40+8 = 48ml squash) and (32+40 = 72ml water). Therefore, glass one has 48ml squash and glass two has 40ml squash.

Question 21: B
B is the main conclusion of the argument. Options A and D both contribute reasons to support the main conclusion of the argument that the HPV vaccination should remain in schools. C is a counter argument, which is a reason given in opposition to the main conclusion. Option E represents a general principle behind the main argument.

Question 22: B
The speed of the bus can be calculated using the relationship: Speed $=\frac{distance}{time}$

$\frac{3 \text{ km}}{0.2 \text{ h}} = 15 \text{ kmh}^{-1}$

The bike speed is therefore ($\frac{4}{5}$ x 15 = 12 kmh^{-1}). Considering that the bus leaves 2 minutes after the bike, it is now possible to write an expression, where d is the distance travelled when the bus overtakes the bike:

$\frac{d \text{ km}}{12 \text{ km/h}} = \frac{1}{30} \text{ h} + \frac{d \text{ km}}{15 \text{ km/h}}$

This expression can be solved by multiplying each term by (12 kmh^{-1} x 15 kmh^{-1}):
15d km = 6 km +12d km
3d km = 6 km
d = 2 km

Therefore, the bus overtakes the bike after travelling 2 km.

Question 23: B
Firstly, determine who will move up to set one. Terry, Bahara, Lucy and Shiv all have attendance over 95%. Alex, Bahara and Lucy all have an average test mark over 92. Terry, Bahara, Lucy and Shiv all have less than 5% homework handed in late. Therefore, Bahara and Lucy will both move up a set. Secondly, determine who will receive a certificate. Terry, Bahara, Lucy and Shiv have absences below 4%. Alex, Bahara and Lucy have an average test mark over 89. Bahara and Shiv have at least 98% homework handed in on time. Therefore, only Bahara will receive a certificate.

Question 24: C

Firstly, construct two algebraic equations: A-18=B-25 and A=$\frac{5}{6}$B

Solve these two equations as simultaneous equations by substituting $\frac{5}{6}$B for A in equation 1:

$\frac{5}{6}$B-18=B-25

7=$\frac{1}{6}$B

B=42

Put B=42 back into equation 2: A= 42 x $\frac{5}{6}$

A=35

Question 25: D

I need to make 48 scones, which makes up 8 batches.
8 batches would take: 35+ 7(25+10) +25 = 305 minutes

I need to make 32 cupcakes, which makes up 4 batches.
4 batches would take: 15+ (4x20) =95 minutes

I need to make 48 cucumber sandwiches
This would take (8x5) = 40 minutes

Adding 305, 95 and 40 minutes is 440 minutes in total. 440 minutes is equivalent to 7 hours and 20 minutes. Adding 7 hours and 20 minutes to 10:45am leads to 6:05pm so I will be finished at 6:05pm.

Question 26: D

The volume of a pyramid is given by the equation:

$v = \frac{a^2 h}{3}$ where v=volume, a=base and h=height

Rearrange to work out the height for each pyramid: $h = \frac{3v}{a^2}$

Pyramid	Base edge (m)	Volume (m³)	Calculation:	Height (m)
1	3	33	$\frac{3 \times 33}{9}$	11
2	4	64	$\frac{3 \times 64}{16}$	12
3	2	8	$\frac{3 \times 8}{4}$	6
4	6	120	$\frac{3 \times 120}{36}$	10
5	2	8	$\frac{3 \times 8}{4}$	6
6	6	120	$\frac{3 \times 120}{36}$	10
7	4	64	$\frac{3 \times 64}{16}$	12

The tallest pyramid is 12m and the smallest is 6m. Subtracting the height of the tallest pyramid from the height of the smallest pyramid leaves 6m.

MOCK PAPER F SECTION 1 — ANSWERS

Question 27: A
Work out the two wages by substituting the information provided into the formula:
Jessica's wage is: 210 + 42 - 3.2 = 248.8
Samira's wage is: 210 + 78 - 8.8 = 279.2

Subtracting 248.8 from 279.2 leave 30.4 so the difference between their wages is £30.4.

Question 28: C
The main conclusion is C. A and B both represent reasons to support the main conclusion of the argument. Option D represents an assumption that is not stated in the argument but is required to support the main conclusion that research universities should strongly support teaching. Option E is a counter argument that provides a reason to oppose the main argument.

Question 29: D
D is the main conclusion of the argument. A is a general principle of the argument, but the argument is more specific to the use of helmets rather than the wider concept of danger in sport and the responsibilities of the governing bodies to sports players. Options B and C are reasons to support the main conclusion. Option E is an intermediate conclusion, which acts as support for the next stage of the argument and as a reason to support the main conclusion.

Question 30: D
There are 10 passengers on the tube at the final stop. At stop 5 there were twice the number of passengers on the tube so 20 passengers were at stop 5. At stop 4, there were $\frac{5}{2}$ times the number of passengers at stop 5 so 50 passengers were present at stop 4. At stop 3, there were $\frac{3}{2}$ times the number of passengers at stop 4 so 75 passengers were on the tube. At stop 2, there were $\frac{6}{5}$ times the number of passengers at stop 3 so 90 passengers were present at stop 2. Similarly, at stop 1, there were $\frac{6}{5}$ times the number of passengers at stop 2 so at the first stop 108 passengers got on the tube.

Question 31: E
- Some students born in winter like English, Art and Music
- There is not enough information to tell whether some students born in spring like both Biology and Maths.
- We don't know what the students born in spring think about Art.
- We don't know what the students born in winter think about Biology.
- There is not enough information to know whether this is true or not.

Subject	Time of Birth		
	Spring	Autumn	Winter
English	Everyone likes	Everyone likes	Everyone likes
Biology	Some like	No one likes	
Art		Everyone likes	Some like
Music			Everyone likes
Maths	Some like		

Question 32: A
The main conclusion is option A that some works of modern art no longer constitute art. B is not an assumption made by the author as the main conclusion does not rely on *all* modern art being ugly to be valid. C is not an assumption because the argument does not rely on artists studying for decades to produce pieces of work that constitute art. This point is simply used to support the main argument. Options D and E are stated in the argument so are not assumptions. A is an assumption because it is required to be true to support the main conclusion but is not explicitly stated in the argument.

MOCK PAPER F SECTION 1 — ANSWERS

Question 33: E
Reducing the price of the sunglasses by 10% is equivalent to multiplying the price by 0.9. the price of the sunglasses is successively reduced by 10% three times and so the price on Monday is 0.9^3 the price of the sunglasses on Friday. 0.9^3 is equal to 0.729 and so the price of the sunglasses on Monday is 72.9% of the price of the sunglasses on Friday.

Question 34: C
It is easier to write out this calculation in the following format:

a b 7 –
   a b
―――――
5 6 5

From the above subtraction it is clear that b must be equal to 2 because 7 minus 2 is equal to 5, which is the unit term of the answer. It is now possible to rewrite the calculation with 2 substituted for b:

a 2 7 –
   a 2
―――――
5 6 5

From the above calculation it is possible to gauge certain facts. A must be greater than 5 because 1 is carried over to the second term:

a ^{1}2 7 –
   a 2
―――――
5 6 5

It is now clear than a must be equal to 6 because 12 minus 6 is equal to 6, which is the tens value of the answer.

Question 35: E
Look at the flat cube net and note the shapes that are adjacent to each other. Sides that are joining on the net will be beside each other on the formed cube. Work through to deduce option E can be formed from the cube net shown.

Question 36: E
The H shape is comprised of 12 squares. The shape's area of 588 can be divided by 12 to give 49, which is the area of each individual square. The square root of 49 is 7 and so the side length of each individual square is 7cm. The perimeter of the shape is comprised of 26 sides and the length of each side is 7 so the perimeter of the shape is 182cm.

Question 37: E
The information provided about the child needs to be inserted into the BMI formula: $BMI = 35 \div 1.2^2$
1.2 squared is equal to 1.44 and it may be easier to work out 3500 divided by 144. The answer needs to be worked out to 3 decimal places for an answer required to 2 decimal places. The answer to 3 decimal places is 24.305 and so the BMI to 2 decimal places is 24.31.

Question 38: C
It is important that the information is inserted into the formula given for calculating the BMR of a woman rather than a man:
BMR = (10 × weight in kg) + (6.25 × height in cm) – (5 × age in years) –161
BMR = (10 × 80) + (6.25 × 170) – (5 × 32) – 161
BMR = 800 + 1062.5 –160 –161

The BMR of the woman in the question is therefore 1541.5 kcal

Question 39: D
This time, the information needs to be inserted into the formula for calculating the BMI of a man:
BMR= (10 x weight in kg) + (6.25 x height in cm) – (5 x age in years) + 5
BMR= (10 x 80) + (6.25 x 170) – (5 x 45) +5
BMR= 800 + 1062.5 -225 +5
The BMR of the man in the question is therefore 1642.5 kcal. The man does little to no exercise each week. It is therefore required to multiply 1642.5 by 1.2, which gives a daily recommended intake of 1971 kcal.

Question 40: B
Slippery Slope describes a series of loosely connected and increasingly worse events that lead to an extreme conclusion. A is not a flaw because the author does not predict a series of undesirable outcomes. C is not a flaw. It is unlikely that correlation has been confused with cause if the American school did not change other aspects of the school day although this is not explicitly stated in the argument. D is not a flaw. A circular argument assumes what it attempts to prove and this is not the case in this argument. E is a counter argument rather than a flaw. B is the flaw in the argument. Just because moving start times later worked in one school in America does not mean that it will work in all other cases.

Question 41: D
Options A and E, if true, would weaken the argument. If the class is more disrupted this will be detrimental to learning, as will less effective teaching. B does not strengthen the main conclusion, which is based on improvement in academic achievement levels rather than activity levels. C does not strengthen the argument as the school curriculum makes no difference to the argument about the science behind teenage brains. If D is true then it suggests that the improvement in grades is a direct effect of the later school starts rather than a mere correlation.

Question 42: B
The main conclusion is that EnergyFirst is expected to expand its customer base at a rate exceeding its competitors in the ensuing months. A does not directly contradict the main argument. It demonstrates a flaw in the argument in that it ignores the fact that other companies may be stronger in other areas and attract customers by other means. However, it does not serve to weaken the main argument. C does not contradict the main conclusion; EnergyFirst could still expand its customer base at the fastest rate even if there is not much competition between energy companies. D would not weaken the argument as it refers to the rate of new customer intake rather than the number of new customers attracted. E, if true, would strengthen the argument because it suggests that visual advertising would attract new customers. B would weaken the main argument because if it were true then investing the most money in advertising would not serve to attract the most customers.

Question 43: A
Option B points out a flaw in the argument, which attributes the healthier circulatory system of vegetarians to diet, but ignores other potential contributory factors to a healthy circulatory system such as exercise. C is not an assumption: the health benefits of a vegetarian and omnivorous diet are not discussed; rather the argument is centred on the negative health ramifications. D is stated in the argument so is not an assumption and option E is a counter argument, not an assumption. Option A is required to support the main conclusion but is not stated in the argument so is an assumption made in the argument.

Question 44: A
First, calculate the number of hours spent flying and waiting. It takes 24 hours in total from Auckland to London, 11.5 hours from London to Calgary and 8 hours from Calgary to Boston. In total this amounts to 43.5 hours of flying and waiting. Boston is 16 hours behind Auckland and so when Sam arrives in Boston it will be 27.5 hours ahead of 10am. The time in Boston will therefore be 13:30 pm.

MOCK PAPER F SECTION 1 — ANSWERS

Question 45: B

This question requires you to find the lowest common multiple. This is the product of the highest power in each prime factor category.

$18 = 3^2 \times 2$ $33 = 3 \times 11$ $27 = 3^3$

Therefore, 3^3, 11 and 2 need to be multiplied together which equals 594 seconds between simultaneous flashes. 5 minutes or 300 seconds needs to be subtracted from 594 in order to find the length of time until the next flash. The time that they will next flash simultaneously is 294 seconds.

Question 46: B

Firstly, calculate the number of students who play each instrument. 21 students play piano, 12 play violin and 3 play saxophone. Point 1 is true because the sum of 21 piano students and 12 violin students is 33, which is 3 more than the total number of students in the class. Therefore, at least 3 students must play both piano and violin. Point 2 is true because only 12 students actually play the violin so there cannot be more than 12 students playing both piano and violin. Point 3 does not have to be true because some of the 9 students that do not play piano may play the violin.

Question 47: D

One way of answering this question is to set out the result after each game:

	Neil	Simon	Lucy
Start	50	50	50
Game 1	100	25	25
Game 2	50	50	50
Game 3	25	25	100
Game 4	12.5	12.5	125
Game 5	6.25	15.625	128.125

After game 5 Lucy has £128.13, however the question is asking how much money Lucy gains. The difference between 128.125 and 50 is 78.13, so Lucy gains £78.13.

Question 48: C

Option A may explain why young drivers are involved in more accidents but does not need to be true for the main conclusion to hold. B would weaken the argument if true as drivers that spend more time driving will have a greater chance of being involved in accidents regardless of age. D is not an assumption, but if true may weaken the argument as it attributes the accidents to unsafe cars rather than unsafe driving. E is irrelevant to the main conclusion: it does not matter whether the young drivers are male or female; arguably steps should still be taken to reduce the number of accidents. Option C represents an assumption that is not stated in the argument but is required to support the main conclusion.

Question 49: D

The total weight of all of the apples is 6 multiplied by 180g, which equals 1080g. The highest value the heaviest apple could take would occur if all of the other 5 apples weighed the same as the lightest apple. 5 multiplied by 167g, the weight of the lightest apple, is 835g. The difference between the weight of all of the apples (1080g) and 835g gives the highest possible weight of the heaviest apple, which is 245g.

Question 50: B

A) True, but not far-reaching enough.
B) Correct answer. Sugar does indeed have an addictive potential as it causes the release of endorphins and the health concerns are well known. This characteristic makes it like alcohol and smoking, and potentially suitable for similar policies.
C) True, but similar to option A) and thus too limited.
D) Potentially true, but also too limited.

Mock Paper F: Section 2

1   *'Social media has risen and fallen'. Discuss.*

Introduction:
- Define briefly what social media entails – e.g. Instagram, Facebook, Twitter, Snapchat, YouTube
- State what can possibly be attributed to the 'rise' and the 'fall' of social media
- Indicate what your overall standpoint is, for example, state why you think social media has risen over the years (increasing number of users, financial success of these social media companies, social media becoming quintessential in our lives) and why there might have been a recent fall (e.g. people eschewing social media platforms, drop in interest over social media platforms)

Rise of social media:
- Exponential increase in the number of social media users over the past ten years
- Social media has become highly influential – e.g. advertisers regularly use social media platforms as a tool to reach their target audience
- Social media platforms have been used for philanthropic and humanitarian purposes – e.g. spreading awareness of certain diseases, plights etc.
- Social media companies have gained significant international clout – e.g. Facebook and Snapchat IPOs
- Social media has becoming a quintessential part of our lives – not only are the younger generation using it, nowadays the older generation have also caught up with the trend and many of them have been using Facebook as a platform to connect with their old friends, colleagues, family members and so on
- There seems to be a platform catered for differing interests and individuals – e.g. Facebook is largely used for staying in touch with friends and relatives, Instagram is for the younger generation and is usually filled with pretty pictures (e.g. holiday pictures, fitness posts, food pictures), whilst Snapchat is predominantly used by teenagers who prefer to chat with each other with the benefit of more privacy
- Social media has also become a lucrative platform – for example certain YouTube stars are able to rake in millions just by posting seemingly regular and mundane videos such as videos of themselves playing games or just chatting about life in general
- Top Instagram stars who have millions of followers are also able to demand premium prices for displaying sponsored products in their posts
- Twitter has also become somewhat of a political tool – as seen from certain politicians such as Donald Trump who uses Twitter frequently as a tool to publish his unbarred thoughts

Fall of social media:
- Increasing criticism over fake news – e.g. Facebook failing to filter out fake news from its News Feed
- Declining relevance of certain platforms such as Snapchat – e.g. fall in share price
- Lack of security – e.g. social media being used by paedophiles to target vulnerable children
- There has been a rapid decline in users in certain platforms – e.g. Facebook has been losing a lot of young users as they perceive Facebook to be increasingly flooded with the older generation and they feel a loss of privacy
- There is a certain level of disenchantment with social media usage – e.g. more and more people are boycotting social media over the fact that it leads to a loss of privacy, creates an unhealthy obsession with portraying a certain image, and it results in a popularity contest at times
- Social media companies have also been criticised for failing to treat the privacy of individuals more seriously – e.g. the data accumulated from a user's social media usage has been used to created targeted advertisements, and such data can be potentially intrusive and include details such as a user's address, ethnicity, sexual orientation, health matters etc.
- Young girls in particular have become more susceptible to eating disorders as a result of the unhealthy body image promoted by social media usage

Conclusion:
- Include a summary of all points given and refer back to your original definition and standpoint

2 Do you think the right to press freedom has become increasingly dangerous and polarising to society?

Introduction:
- Briefly state the notion of the right to press freedom and why it is protected
- Explain why this right might have become increasingly dangerous and polarising to society
- State your standpoint
- For example, you should mention why right to press freedom has traditionally been protected – e.g. to preserve journalistic integrity and to ensure that the people have a right to know about what is going on (such as government corruption)
- You should then go on to talk about how this right might have manifested over the years and has been abused – for example irresponsible and unscrupulous journalists who create sensationalist and fake news articles in order to generate more readership

Possible reasons for why it has become increasingly dangerous and polarising:
- Lack of effective quality check – e.g. rise of fake news
- Irresponsible journalism leading to the public being misled – e.g. bad reporting during the Brexit campaign
- Sensationalist articles polarising society – e.g. articles bashing migrants
- Far-right activists inciting hate by delivering controversial speeches
- Used as a tool for slandering and defamation of politicians
- Right-wing news articles such as The Sun and The Daily Mail are particular notorious for focusing on gossip and publishing news articles in a way that will increase their readership – e.g. sensationalist headlines, focusing on stereotypes, pandering to the right-wing community by publishing opinions that come across as intolerant and insensitive
- On the flip side of the coin, left-wing newspapers have also been known to public news articles that are heavily-skewed and polarising – e.g. The Guardian publishing news articles that are very liberal and insistent on certain topics which might not be agreeable to a more conservative group of people
- Journalists have been heavily squeezed by their profit margin and they can only survive by increasing their readership, and may have to resort to focusing on sensationalist news even if it goes against their professional integrity – e.g. The Sun and The Daily Mail have a much higher readership than considerably more neutral or high-brow newspapers such as The Times or The Telegraph
- People choose to read what aligns with their views so a reader will only read right-wing newspapers and reinforce their pre-conceived notions, and the same goes for left-wing readers who will only actively read left-wing newspapers – this results in society becoming increasingly polarised and people on both sides of the spectrum become more extreme and intolerant of alternative viewpoints

Possible arguments against:
- Press freedom crucial in ensuring an effective check and balance against the misuse of power
- Crucial in exposing scandals
- Press freedom has been pivotal in the recent Harvey Weinstein scandal and the #MeToo movement
- Public has a right to unhindered access of information in order to form their own informed decision
- There is a plethora of alternative news articles available in the market – instead of restricting press freedom we should encourage people to read up more widely and read from different sources in order to come to a more sensible and informed viewpoint
- A free and independent journalism industry is key to ensuring that we are a functioning and effective democracy by ensuring that the population is always kept informed about what is happening in the country and in the world, and bad policies and scandals cannot be swept under the rug easily
- Without press freedom, many controversial incidents would not have come to light, such as the Panama and Paradise Papers

Conclusion:
- Include a summary of all points made, and give a balanced overview of both sides of the argument.
- Reiterate what your overall conclusion is and why you have arrived at that conclusion
- As always, if you have a strong opinion, make sure you defend it well and state in the conclusion why the arguments or counter-arguments you have raised made you come to a certain conclusion

3. *'Artificial intelligence replaces humans and leads to unemployment'. Do you agree with this statement?*

Introduction:
- Define artificial intelligence
- Give possible examples of the use of artificial intelligence – e.g. driverless cars, automation
- Explain why you think this might or might not lead to unemployment
- For example, you can mention how artificial intelligence has already been slowly phased out in certain industries – e.g. replacing call centres, the use of artificial intelligence in smartphones and even self-checkout machines have been replacing cashiers

Possible arguments for artificial intelligence:
- Artificial intelligence supports the economy rather than compromising it
- Artificial intelligence leads to greater efficiency
- Artificial intelligence reduces human error
- Artificial intelligence creates more high-end jobs that require more skills such as coding and engineering
- Mundane jobs that are low-paid and menial will be replaced with high quality jobs
- Artificial intelligence is highly complementary to human labour as it reduces the need for humans to do repetitive and menial chores and allows workers to focus on more complex and sophisticated manners
- Artificial intelligence has a wide application and can be used for anything such as self-checkout machines to being employed in complex, legal transactions by automating certain processes
- The rise of artificial intelligence has led to the boom of our tech industry and has created many relevant jobs in the field of software engineering and coding
- Artificial intelligence has solved many problems such as a declining population in certain countries such as Japan which necessitates artificial intelligence in some traditionally labour-intensive industries
- Artificial intelligence has also made it possible for the economy to be more efficient, such as allowing driverless trains to operate and hence allowing for a 24-hour service
- Artificial intelligence has also been used in high-end medical technology due to its much greater precision, helping to improve healthcare

Possible arguments against artificial intelligence:
- Artificial intelligence disproportionately affects low-income groups as their jobs are the most susceptible to automation
- Artificial intelligence might result in a lack of human touch – e.g. replacing drivers and waiters
- Artificial intelligence is potentially unreliable and unstable
- Certain jobs are not amenable to artificial intelligence – e.g. jobs requiring communication and interaction
- The cliché of artificial intelligence creating super robots that take over humankind may have some truth to it as robots become more sophisticated and are able to mimic human actions more – there is a limit as to where our ethical boundaries are and to what extent should we make such robots humanlike
- Whilst artificial intelligence is promising and seems to be able to develop our economy, it is still largely uncharted territory and it is questionable whether artificial intelligence can truly replace certain skilled jobs such as lawyers and doctors which require years of training
- Automation results in a massive cut in labour-intensive jobs such as manufacturing and production – these workers may not always be able to upgrade their skills to keep up with the changing times and they may become unemployed as a result
- People are resistant to change and certain group of workers may be heavily protected by trade unions and refuse to lose their jobs to automation – e.g. train drivers

Conclusion:
- Include a summary of all points made, and give a balanced overview of both sides of the argument.
- Reiterate your overall conclusion and explain how you arrived at your conclusion
- Come up with a sensible conclusion based on the arguments and counter-arguments you have raised, and briefly explain why you have arrived at this conclusion

4 Does our society place too much emphasis on a university education?

Introduction:

- Briefly state why a university education is emphasised in our society
- State the pros and cons of university education
- State your standpoint as to whether we truly over-emphasise university education
- For example, you can state how many students enrol in subpar universities that do not truly improve their job prospects or do not end up learning anything useful in university due to the lack of proper teaching
- The example of the Blair administration wanting to raise the number of university graduates to 50% of the cohort can also be used to show how this might result in an unsustainable number of university graduates in the job market, leading to a mismatch in demand and supply of certain jobs

Arguments for university education:

- A higher level of education is beneficial to society as a whole
- Keeping students in education will prevent them from being bored and going astray
- University education inculcates useful skills such as teamwork and communication
- University is a good period of time for students to form lasting bonds and networks
- University education is an important sector of our economy
- There is a positive correlation between higher levels of education and future earning power
- University is not solely about academics, even though it forms a huge factor, but there are also several extra-curricular activities that students tend to engage in, such as sports and clubs, which help to develop their leadership and organisational skills
- Arguably being in higher education is a good way of giving students time to properly think about what they want to do in life later on, giving them a safe space to develop their own perspectives and thoughts without the consequences of the real world, and helping them to mature as adults upon graduation
- Being in university also means that you will be part of a network of alumni, which is highly beneficial in the working world when you establish professional relationships and being part of an alumni network is useful for forging connections
- Certain important skills such as effective writing, communication, presentational skills, computer skills and editing skills can be picked up at university which is highly beneficial for working life
- Not all students go into university purely for the sake of getting a good job – some might genuinely enjoy the intellectual stimulation, the friendships that can be made, living in a hall together, or simply if they are interested in academia

Arguments against university education:

- Not all students are academically suitable for university education
- Some universities do not provide a high quality of education
- Many students incur enormous debts from university and are unable to secure well-paying jobs
- Some students waste their time in universities by not focusing on their studies and spend their time partying and drinking instead
- Universities do not always produce graduates that are suitable for the demands of the workforce
- University education has become a huge burden on taxpayers – many student loans do not get paid off fully because many students do not end up with jobs that are well-paying upon graduation
- Universities have also led to a segregation in certain cities – for example the town vs. gown culture in Oxford and Cambridge, and also the increasing gentrification as students dominate certain towns
- Teaching in certain universities have been known to be subpar, and this affects even certain top universities, as professors prefer to focus on their research at the expense of providing high quality teaching to undergraduate students – this is partly due to the fact that university rankings place an emphasis on research quality over student satisfaction
- Universities have become less selective in accepting students and have a tendency to accept a large number of students because they are fee-paying and this results in a larger class size

Conclusion:

- Summarize the key ideas explored previously – the ideas associated with university education and whether we should decrease our emphasis on it
- Ensure to link back to the original statement – should less emphasis be placed on university education?

Final Advice

Arrive well rested, well fed and well hydrated

The TSA is an intensive test, so make sure you're ready for it. Ensure you get a good night's sleep before the exam (there is little point cramming) and don't miss breakfast. If you're taking water into the exam then make sure you've been to the toilet before so you don't have to leave during the exam. Make sure you're well rested and fed in order to be at your best!

Move on

If you're struggling, move on. Every question has equal weighting and there is no negative marking. In the time it takes to answer on hard question, you could gain three times the marks by answering the easier ones. Be smart to score points- especially in section 2 where some questions are far easier than others.

Make Notes on your Essay

You may be asked questions on your essay at the interview which might take place upto six weeks after the test. This means that you **MUST** make short notes on the essay title and your main arguments after you finish the exam. This is especially important if you're applying to Economics or PPE where the essay is discussed more frequently.

Afterword

Remember that the route to a high score is your approach and practice. Don't fall into the trap that *"you can't prepare for the TSA"*– this could not be further from the truth. With knowledge of the test, some useful time-saving techniques and plenty of practice you can dramatically boost your score.

Work hard, never give up and do yourself justice.

Good luck!

Acknowledgements

I would like to express my sincerest thanks to the many people who helped make this book possible, especially the 15 Oxbridge Tutors who shared their expertise in compiling the huge number of questions and answers.

Rohan

About Us

Infinity Books is the publishing division of *Infinity Education Ltd*. We currently publish over 85 titles across a range of subject areas – covering specialised admissions tests, examination techniques, personal statement guides, plus everything else you need to improve your chances of getting on to competitive courses such as medicine and law, as well as into universities such as Oxford and Cambridge.

Outside of publishing we also operate a highly successful tuition division, called UniAdmissions. This company was founded in 2013 by Dr Rohan Agarwal and Dr David Salt, both Cambridge Medical graduates with several years of tutoring experience. Since then, every year, hundreds of applicants and schools work with us on our programmes. Through the programmes we offer, we deliver expert tuition, exclusive course places, online courses, best-selling textbooks and much more.

With a team of over 1,000 Oxbridge tutors and a proven track record, UniAdmissions have quickly become the UK's number one admissions company.

Visit and engage with us at:
Website (Infinity Books): www.infinitybooks.co.uk
Website (UniAdmissions): www.uniadmissions.co.uk
Facebook: www.facebook.com/uniadmissionsuk
Twitter: @infinitybooks7

Printed in Great Britain
by Amazon